# FOREWORD

While it is one thing for historians to go over fairly familiar ground over and over again, [illegible] another to delve really deep and come up with page after page of hard-won and revealing original research, especially when this is backed up by previously-unpublished photographs from such untapped sources as old and treasured family albums. To do all this and make the end result not only deeply factual but also highly enjoyable to read – even amusing in places – would take a special kind of skill even if the author were writing about the history of a single firm. To write so flowingly about two firms - as my dear friend Pat Freeman has ably done on this occasion – is even more praiseworthy.

Pat's book about two East Anglian family firms, W.G. Smoothy & Son of Rochford, Essex and M. Bloomfield & Sons of Debenham, Suffolk, more than keeps up the standard of titles about aspects of the past previously published by the East Anglian Traction Engine Society, and I am delighted to see this fine tradition continuing to enrich the store of regional knowledge for both ourselves and generations to come.

Against a broad backcloth of history going back to Roman times, Pat tells how the seeds of these two very different firms – one specialising in ploughing and road making contracting and the other in buying and selling – came about and how each firm developed over the years. While discovering intricate details about the firms' founders, their families and employees, and the steam and associated machinery with which they dealt, Pat almost unwittingly builds up a fine picture of general history, ranging from the development of road-making to some revealing facts about the fortunes of Garrett's of Leiston, Suffolk and the effect the aftermath of the Irish Potato Famine had on a Quaker by the name of John Fowler. Not a stone has been left unturned in Pat Freeman's passionate search through the flotsam and jetsam of regional history, and he's done both the Society and everyone fortunate enough to pick up a copy of this book proud.

It has been a privilege to have been asked to write this foreword, and I am proud to have a writer of such talent on *Old Glory's* team of contributors.

Peter Kelly.

PETER KELLY
Editor, Old Glory Magazine.

# CONTRACTOR
# and
# FACTOR


*ISBN 0 9523111 5 1*

*Typeset and Printed by:*
**PETER REED PRINTERS LTD**
Chelmsford, Essex
Tel: 01245 474080

*Published by the*
**EAST ANGLIAN TRACTION ENGINE CLUB**
*for the*
**EAST ANGLIAN TRACTION ENGINE SOCIETY**
Registered Charity No. 296027

*Other books published by the Club:*

THE OLD GUARD
THRESHERS YARD
BRICKS & ROLLERS
FORTY YEARS ON
THE HEAVY GANG
THRESHING & THATCHING
THREE COUNTIES STEAM TOUR

"VAPOR OMNIA VINCIT"

"VAPOR OMNIA VINCIT"

# CONTENTS

*This Garrett 'Suffolk Punch' W.No.33180, built in 1919, was owned by John Goddard, of Tunstall, who had it maintained by Bloomfields during its working days. This engine now belongs to Leiston Long Shop Museum.* Keith Bloomfield Collection.

# ACKNOWLEDGEMENTS

I would like to thank wholeheartedly the many individuals who have contributed verbal or written information or photographs which have enabled me to write this book and I beg forgiveness of the many if I name but few.

John Smoothy, a member of the East Anglian Traction Engine Club, was not only the catalyst for writing this book but he also gave me a great deal of help with its preparation. John is a great grandson of W.G. Smoothy and as such he was researching the family history when, at some stage, he spoke to me about the steam engines that the firm had owned. After chatting with him I saw the potential for another publication in the East Anglian Traction Engine Society's series of local history books and the committee gave me the go-ahead to write it. John is a proficient typist and he volunteered to type my hand-written manuscripts and also to get the information onto floppy discs as required by the printer. Thank you, John.

Another great grandson of W.G. Smoothy is Derrick Smoothy and I am pleased that he also provided me, as had John, with details of the family history and photographs as well as copies of the firm's Engine Register.

Whilst writing the book the E.A.T.E.S. committee suggested that I include the history of M. Bloomfield & Sons and this I was happy to do.

I therefore thank two of Malcolm's sons, Keith and Neil, for all the help, information and photographs that they gave me.

I am also grateful to Robert Oliver, Chairman of the Steam Plough Club and to Lyndon Shearman, ex-President of the Road Roller Association, for their contributions to the Chapters on Road Making and Steam Ploughing.

Finally I would like to thank Peter Kelly, Editor of "Old Glory", for kindly agreeing to write the Foreword.

Thank you all

Pat Freeman

*Taken in Debenham High Street, believed about the time of the first World War. The steam roller carries the pre-registration County Plate of 'Essex CC No.312'. It was once owned by Smoothys who acquired it on February 17th 1903 from George Porter (or Parker – the engine register entry is indistinct) of Hackney. It is an Aveling & Porter roller W.No.2221, built in 1886. Smoothys gave it Fleet No.7 and whether or not it was still in their possession when this photograph was taken is not known. Smoothys did do work in Suffolk and still had the roller in August 1913 when they fitted a new fire box to it in their works.* Neil Bloomfield Collection.

# INTRODUCTION

During the last few years the East Anglian Traction Engine Society has published several books on aspects of the past, centring on small firms which have made their contribution to East Anglian local and industrial history.

The contents of these books have mainly been obtained from the relevant family members, many being of advanced years, and it has been satisfying and informative to commit their memories to paper.

It is one of the aims of the Society to continue this series of publications and I am pleased to have added "Contractor and Factor" to those books previously written.

As befits a Society with the words "Traction Engine" in its title, all of the books in the series have featured some aspects of steam engines and their owners and drivers and "Contractor and Factor" follows the same format.

Two family firms take pride of place in the book: W.G. Smoothy & Son of Rochford, Essex, which ceased trading in 1946 and M. Bloomfield & Sons of Debenham, Suffolk, which is still active.

During the first half of the 20th century both firms were very much involved with steam engines and allied farm equipment, although their areas of operation were somewhat different. W.G. Smoothy's main emphasis was on the business of Contracting, both with steam rollers and steam ploughs, whilst M. Bloomfield concentrated on Factoring, i.e. buying and selling on their own behalf and through agencies.

Initially both firms had a sole founder and then, mainly with the help of family members, they blossomed out such that their businesses met the needs of the local communities wherein they were situated.

The opening chapters of "Contractor and Factor" describe, in not too much detail, aspects of traditional road making and steam ploughing which have relevance to the contents in the rest of the book.

Almost a hundred photographs, suitably captioned, have been used to supplement the text and the majority have been obtained from family albums. It will therefore be the first time that the general public have been able to view these extremely interesting records.

Pat Freeman

C.P. Freeman, C.Eng.M.I.E.E.

# Road Making - Steam Roller Contracting

When the Romans departed from our shores in the 5th century AD, they left behind a network of about 10,000 miles of very well-made roads. They were able to construct these because they had acquired the necessary expertise, and evolved the tools needed, over many centuries of road building. They commenced with the infrastructure required in Rome, an example of which was the paved Appian Way, and had continued with extensive road networks needed for troop movements as their military might spread upwards through Europe into Britain. When they left in 410 AD, they took with them, crucially, their knowledge and experience of road building and road maintenance. The result was that between 410 AD, and the first Turnpike Act in 1663 it is almost certain that no true roads were built and the vast majority of the Roman roads had either fallen into disrepair or had vanished.

The making of a Turnpike road, however, left a lot to be desired as it only consisted of roughly levelling out the line of the new road, removing a little topsoil, then dumping down stones of different sizes and allowing them to be consolidated, as best they could, by passing traffic.

*Aveling & Porter steam roller fleet No.23 HJ1732.*

Derrick Smoothy Collection

The history of modern road-making can be traced back to the 18th. century when the French Government set up the École des Ponts et Chaussées (School of Bridges and Roads) to teach road building techniques for military purposes. To do this the School authorities studied Roman methods which were, in turn, studied by British road builders, notably Metcalfe and Telford.

The basis for a good road is

adequate consolidation of the materials used and there is no doubt that the design and development of the Steam Roller was a major breakthrough for this operation.

The first Steam Road Roller was designed by the Frenchman Monsieur Louis Lemoine, of Bordeaux, who took out a patent for this type of machine in 1859 and demonstrated it in operation in 1860. A few years later British steam rollers were being designed and manufactured, and their usefulness was quickly appreciated by those authorities that were involved in road building and repair. However, the cost of the rollers was such that many of these could not afford to own their own machines, so they had to rely on hiring them from entrepreneurs who had spotted the niche in the market and had established contracting businesses to cater for the demand.

By the 20th. century a well developed system of operation had developed and road contractors could be broadly divided into two classes, the road builders and the surfacing contractors. The former were those who excavated the bed of the road, laid drains, kerbs, and the broken stone sub-base. Their work consisted of building new roads or widening, or reconstructing, existing roads or putting in roads on housing or industrial estates. The roadmakers were not, as a rule, general builders. There were exceptions, of course, and

*Aveling & Porter steam roller was fleet No.34 when at Smoothys. Reg No. HJ5751. Taken in Jack Shuttlewood's yard, near Braintree.*

Arthur Clarke Collection

*Aveling & Porter steam roller Fleet No.12. Taken at South Benfleet in 1908. Oswald Smoothy is at the controls.*

John Smoothy Collection.

*Fowler steam roller Reg No. NY2227 Fleet No.26. Taken at Lambeth on August 3rd 1950 when owned by Percy Parsons of Cheam.*

John H. Meredith Collection.

THE COLEMAN
FLAPPER
SPRAYING MACHINE
We claim that this machine is without a serious rival. Every feature is a specialised job based upon long experience of the work for which the machine is designed. Simplicity of operation is ensured, and with ordinary care the machine will continue to function satisfactorily, and give perfect results over a long period of years.
Uniform distribution.
Easy variations for widths and quantities.
Rapid and economical working.
Simplicity of operation.
Low cost of upkeep.
Solid rubber tyres if required.
Shafts or drawbar optional.
MAINTENANCE
The maintenance costs of a Flapper are negligible, for there are no fragile or intricate parts likely to wear out quickly or cause trouble and delay by breakdown.
Operation is of the simplest possible nature, and the inexperienced user can without difficulty obtain results which will give unqualified satisfaction to the severest critic. The "Flapper" is particularly economical in use, and will moreover enable you to complete your spraying programme in the shortest possible time.
'PHONE: 45519
THOMAS COLEMAN & SONS LTD • DERBY
GRAMS: "ROADS"

Wimpeys is an example of an house builder who was competent at road making. Roadmakers varied in size from very large civil engineering firms to small family businesses.

Road surfacing specialists could be divided broadly into three groups, the Tarmacadam firms, the Asphalt firms and the Tarsprayers.

In the first group, some had their own quarries and mixing plants for making tarmacadam, whilst others bought it in and laid it. Much work was carried out for Councils on public roads, but there was also a useful trade surfacing factory and housing estate roads and, at the lower end of the scale, footpaths, school playgrounds and drives. In the early years of the 20th century there was some mixing of tarmacadam on site, with tar boilers to melt the tar and concrete mixers to mix the macadam; this was usually done on smaller jobs. Slightly up the scale mixing plants would be set up, often driven by a Portable steam engine, alongside major road schemes. However, much tarmacadam was mixed at the quarry and transported by rail to be unloaded by hand, often being delivered to site by steam tractors and trailers from the nearest railway siding. Many tarmacadam companies were set up by former employees of other firms, so that a small area might contain a proliferation of firms whose histories were interconnected by the movement of employees between them.

*Glossop's Tar Pot in Old Church Yard, Romford, July 1936.*

John Smoothy Collection.

Asphalt companies were the Rolls Royce's of the Trade. Hot rolled asphalt was the preserve of the few, being expensive to make and lay. Failure was costly for the contractor and high standards of quality control and laboratory supervision were necessary.

Over the years the distinction between tarmacadam and asphalt companies has become blurred and now most surfacing firms will take on either kind of work.

The tarspraying contractors worked in the summer months only, spraying the roads with tar, covering them with chippings, then rolling them. The spraying was mechanised very early on

*A typical early road-making scene, put on by members of the Road Roller Association*

R.R.A. Archives

*Fowler steam roller - Tar Sprayer, Works No. 17506, owned by Andy Filewood.*

Photo: Pat Freeman

(though odd pockets of hand spraying persisted until the 1970's), but for a long time the chippings were spread by hand. Casual labour was used, working long hours but earning good money and then usually passing most of it over the bar in the local pub. The work was fast and furious, with roads remaining open to traffic whilst the work progressed. Because tar will not adhere to a wet road, work stopped if it rained - or did not start, if it looked like rain or had rained recently (and sometimes if the foreman had a good thirst when the pubs opened!). Consequently there was always a mad rush to finish off all that season's work before the Autumn weather turned cool.

All the different kinds of jobs mentioned above, undertaken by road makers or road surfacers, provided work for rollers.

*Wallis & Steevens, ex-Smoothy fleet No.32. Charlie Taylor is on the right.*

Photo: Kevin Swann.

To meet this demand there came into being firms who specialised in hiring out rollers and also threshing firms who added roller hiring to their existing business, quite often by using convertible traction engines for both purposes.

Some of these firms were very large, such as the Eddison Steam Rolling Co. with, at its peak, some 400 steam rollers in operation, but many were quite small, such as Smoothys with a dozen or so machines at any one time; there were also a lot of owner-drivers. Much work was initially carried out for Local Councils who could not easily afford to buy their own rollers.

There was also a useful demand from building contractors who either did not have their own machines or who hired them to cover peak demands. Rolling for differing kinds of work and surfaces meant that the roller driver had to choose the correct technique for the particular job in hand.

Rolling broken stone was steady work, hard for neither roller nor driver, and one could spend days rolling the same bit of road until it was absolutely solid. It was a question of keeping the roller moving all day long and not getting too close to newly-laid kerbs. Tarmacadam required more skill and a roller that was in good order. Here the roller driver could make or ruin the road by his technique as, in addition to preserving the camber and understanding how tarmacadam lost its plasticity as it cooled, the roller driver had to be able to roll right alongside kerbs. Rolling asphalt was similar, but most asphalters had their own rollers and drivers who specialised in this type of work. However, the road rolling contractor's driver had to be prepared to tackle hot-rolled asphalt if the job came his way.

*Sentinel Tar Sprayer No.8666 built in 1932 as a DG4 tipper. Acquired by Glossops in 1952. This was the last steam sprayer owned by Glossops and in 1977 it was reconditioned as an example of the old type sprayer in its working livery. It is now owned by Coppard Plant Hire Ltd of Crowborough, Sussex.* Photo: Pat Freeman.

Rolling chippings for tarsprayers, or "chip-chasing", was a frantic business, and often two rollers were used. Everyone was going flat out and the roller driver simply had to keep up. Rolling chippings into the tar required little skill and some fairly decrepit rollers worked out their time on this sort of work. However, the roller-driver had to be adept at manoeuvring his roller at speed and he was often one of the most skilled men in the happy-go-lucky gang (and also one of the few who also worked in the winter).

One other kind of work deserves a mention. Councils had their own gangs for patching work, road improvements and the like and though many had their own rollers they often hired extra rollers to work with their gangs. This was good, steady, work for rollermen as patching was not demanding (although scarifying could be) and the pace of work was generally somewhat slower than in the contracting sector.

It is worth mentioning that the earlier, generally rural, water-bound roads required the services of the steam roller and driver, mostly during the winter months.

These roads were covered with a layer of broken stones, flints or granite chips which were then rolled down hard. A film of hoggin, sharp sand or even road sweepings was laid on top and thoroughly watered from a water cart fitted with a spraying device. The roller driver would then quickly work his machine back and forth, starting from the side of the road, "half-wheeling" about 6 inches at a time, to the centre of the road. He would then repeat this operation from the other side of the road. Whilst this was being carried out, men with stiff brooms would sweep the top film into the road material as the roller passed to and fro until a thick slurry was produced. This was left to dry-out and the surface would then be sufficiently hard to cope with the traffic of that time.

*Aveling & Porter steam roller Reg No NO3326 Fleet No.29. Taken May 14th 1929.*

John Smoothy Collection.

Steam roller contracting began to die out in the 1950's mainly due to the rapid influx of diesel machines with their inherent advantages (light weight for hot-rolled asphalt, ease of starting for daily works schedule etc.) and to the fact that a lot of steam rollers were getting to the end of their useful life and the necessary skilled steam engine drivers were difficult to obtain. The decades of 1950-1970 therefore saw the demise of almost all the old steam roller contracting firms, including Smoothy's of Rochford who were one of the earlier fatalities.

There are now no businesses whose sole income is derived from steam roller hiring, the last of which was probably Lyndon Shearman's, a past President of the Road Roller Association, who finished operating in 1984.

*Two Aveling & Porter steam rollers. At the front Fleet No.29, NO3326, at the rear XD8056 which does not appear in Smoothy's Engine Register or R.L.S. list. Taken at Hill Hall, Theydon Mount, May 1927.*

John Smoothy Collection.

# Roller Types

Over the years much effort has been expended on designing and developing rollers to meet the needs of specific types of surface work.

The most common form of road roller, whether steam, petrol or diesel, was always the traditional three-wheel roller, with its broad front roll (actually two rolls side by side for ease of turning) and high, narrow, hind rolls. Contrary to what might be expected, it is the hind rolls which do most of the work, for they carry most of the weight and it is concentrated on to the narrow width of the rolls. The three-wheel roller was ideal for the old water-bound stone roads and its narrow rear rolls were excellent for punching down the rough stone foundation, but its uneven weight distribution was a disadvantage on tarmacadam or asphalt.

Because of this, so as to give an even weight distribution, the Tandem roller was developed for use on tarmacadam and asphalt. Although pioneering work had been done earlier, it was in 1898 that Thomas Green & Sons produced the first of a new generation of Tandems for surface work. They were mechanically similar to the three-wheel steam rollers, but with a single wide rear roll instead of the narrow rolls. The even weight distribution was helpful, but when the roller reversed it could leave small depressions in the surface every time it did so.

*Barford & Perkins diesel roller. Taken at Hornchurch, June 1936.*

John Smoothy Collection.

Quick reverse Tandems were therefore developed and in 1911 Aveling & Porter built one with a side-slung engine affixed to the right hand horn plate, after the manner of a Shay locomotive, enabling it to be geared directly to the driving roll. An example of such a Tandem roller, still in steam, is Works No. 7411 which was originally supplied to the London Borough of Fulham in 1911.

Only nine of this type were made and No. 7411 is the only survivor.

However, an acceptable solution to the quick-reverse problem was not achieved in mass production form until 1923 when Wallis and Steevens brought out their "Advance" three-wheel roller which was suitable for both foundation work and high class surfacing. This design was the forerunner of all subsequent three-wheel rollers, steam and diesel alike. It had a twin cylinder high pressure engine, with no flywheel, and it was considered by the makers to have five principal features which made it eminently suitable for surfacing work. These were:

(1) cambering rear axle to allow the hind rolls to adjust to the camber of the road,

(2) even weight distribution, with its wide rear rolls,

(3) quick reverse because of the type of engine used and the absence of a flywheel,

(4) direct-acting steering which was more accurate than the old worm and chain method, and

(5) clear-sided, i.e. no part of the roller or scarifier projected beyond the edges of the rear rolls, so it could roll right up to kerbs, walls, lamp-posts and the like.

To try to further improve the roller's surfacing capabilities Tri-tandems were produced, but they were never popular in the rolling trade.

*Barford & Perkins roller. Taken at Southend, April 1935.* John Smoothy Collection.

# STEAM PLOUGHING

## *It's Birth*

At the turn of the 19th. Century the population of England was beginning to increase rapidly and great efforts were being made to expand food production to satisfy the needs of the extra numbers of mouths to be fed.

Tillage of the soil, at that time, was by horse-drawn plough and it was a slow, labour intensive, process such that not much more than an acre per day could be ploughed. Farmers wished to increase this throughput, not only to meet the needs of the expanding population, but also to reduce expenses and increase profits.

*No.6 set of ploughing engines on Foulness Sands, August 2nd 1922.*

John Smoothy Collection.

*No.6 set of ploughing tackle. October 7th 1920. The rear engine was normally driven by Norman Smoothy.*

John Smoothy Collection.

*Fowler Ploughing engines No.2 set. Nearest is Reg No. NO4920*

Derrick Smoothy Collection.

Efforts were therefore made to find alternative methods of turning over the soil and a breakthrough came in 1832 when John Heathcoat and Josiah Parkes patented what was probably the earliest steam plough. By 1835, this was capable of ploughing an acre of land to a depth of nine inches in just under two hours.

In 1839 "The Royal Agricultural Society of England" was formed to organise shows and exhibitions, at different venues throughout the country, which would provide an opportunity for inventors and manufacturers to display their designs and products in working condition to the farming world.

This was the stimulus for many people to come up with ideas for systems and machines to improve agricultural output and perhaps the best known name attributed to the most important advances in this area was that of Fowler; John Fowler in particular.

*Fowler Ploughing engine Reg No. HK9842 Fleet No.3. Norman Smoothy is steering.*

Derrick Smoothy Collection.

He was a young Quaker, working for a railway engineering firm, and when he was on holiday in Ireland in 1849 he was horrified at the distress caused to the population by the failure of the potato crop in 1846 and 1847.

He thought that if the fields in Ireland could be properly drained it would greatly assist the potato crops to flourish. He felt so strongly about what he had seen that he decided to leave his railway job and concentrate on improving methods of drainage and cultivation of the soil.

A mole plough, or mole drainer, was already a practical proposition for deep drainage. It consisted of a flat metal blade that could be pulled vertically through the soil to a depth of about three feet. At its lower end it had a bullet-shaped nose which, when the blade knifed its way through the ground, opened up a continuous 'mole hole' into which excess water could seep, then drain away to the field edges and into a ditch. John Fowler designed and patented a windlass-drawn mole plough, which was made by Ransomes of Ipswich and exhibited at the Royal Show, held at Exeter in 1850. However, this had the added feature of being able to lay a continuous line of clay drain pipes into its 'mole hole', thereby making a more efficient and longer lasting drainage system.

Its drawback, as with all previous models, was that the mole plough had to be hauled back and forth across the field by horse or man power so, in 1851, John Fowler used a Portable steam engine to do this slow and heavy work. The success of this encouraged him to concentrate on the design and construction of cable-drawn steam-operated ploughs.

In 1856 Ransomes built for Fowler a special Portable engine with an integral double-drum windlass which worked with a movable anchor situated on the other side of the field. This method of operation was tested at Nacton, Suffolk and achieved the target rate of ploughing one acre in one hour.

Following on from this Fowler patented a system of ploughing with two engines, one on either side of the field.

By the early 1860's the system of using double-engine tackle in which two self-moving engines, each with its own winding gear, pulled a plough back and forth across the field became pre-eminent and the credit for this was due, in large part, to the efforts of John Fowler. Unfortunately he was killed in a hunting accident in 1864, but his brothers and their heirs carried on to successfully develop the business such that the name of Fowler, to this day, is synonymous with the Steam ploughing engine.

## *It's Operation*

A spectator at one of today's many rallies, who watches a demonstration of steam ploughing, will be witnessing a method of operation that has changed little since its application well over 100 years ago.

There are essentially two methods of setting-up a two-engine system of ploughing, depending on individual choice. The method described below was probably used by working ploughing gangs as it is the quicker of the two, and in those days time meant money to the ploughmen who were often paid an acreage bonus.

Initially two ploughing engines, A and B, arrive at the chosen field, with A towing the plough. The wire drum-rope of B is then connected to the axle of the plough and A proceeds to the other side of the field towing the plough, thereby also pulling along B's rope. When at the far side, A's rope is disconnected and it takes up its working position, then its rope is coupled to the plough axle which is now connected to the drum-ropes of both A and B and is ready to be hauled back and forth across the field.

*Fowler Ploughing engines. Left is HK9841 Fleet No.7, right is HK9840 Fleet No.5.*

D.C. Letch Collection.

After each haul the non-hauling engine will move along the headland the width of the plough's work so that it is in position for the next haul. On large fields, in hilly country or in bad visibility the engine drivers may not be able to see each other so they operate a system, using their steam whistles, to control the sequence of pulls and this works effectively.

A typical 'gang' to operate this two-engine system would comprise of a Foreman, two drivers, a plough or cultivator steersman and a cook-boy and typically the equipment would consist of two steam ploughing engines, a double ended 6/8 furrow plough, a cultivator, a water cart and a living van large enough to house the gang.

A long day was worked, often from dawn until dusk, and the gang would be expected to plough between 80 and 100 acres a week.

Gradually more and more farmers recognised the advantage of steam ploughing, but the price of the engine and tackle was way out of reach of all but the wealthiest. This led to Contractors appearing who purchased sets of engines then obtained contracts from farmers in their area to do their soil cultivation.

Farmers often helped things along by cutting down and grubbing up hedges and filling in ditches to give the engines more room to manoeuvre and allow longer hauls. Some farmers even made ponds close to where engines would work to save longer journeys for the water carts.

One of the most famous steam plough contractors was Ward & Dale of Sleaford, Lincolnshire, who operated at one stage twenty four double-engine sets of ploughing tackle. On the other hand there were many firms like W.G. Smoothy, of Rochford, Essex, who only had a few sets.

*Fowler Ploughing engine Reg No. HK9841 Fleet No.7. Taken at Read's of Hockley January 1935.*

John Smoothy Collection.

# *It's Demise*

Towards the end of the 19th. century farming went into depression with the knock-on effect of reducing the demand for steam ploughing. These were hard times up to the outbreak of the First World War when the threat of a German submarine blockage meant that food production had to increase to replace what was normally obtained from overseas.

Thousands of men and horses were taken from the land for military purposes and farmers had to rely heavily on steam ploughing and direct ploughing by tractors. The Government did order 65 sets of ploughing tackle from Fowlers but did not place the order until 1917. Meanwhile, to make up for the shortage of steam power, many tractors had to be imported from America. For example, the first 7000 Fordson Model F Tractors that came off the production line in 1917/18 came to the UK and many farmers were able to gauge their capabilities for ploughing purposes.

After the war the 65 Fowler sets were released onto the open market and were sold directly to farmers and steam ploughing contractors, thus reducing the demand for new engines directly from the manufacturers which caused them concern. In fact Aveling & Porter stopped making steam ploughing engines in 1921.

Many farmers had obtained experience of direct ploughing by tractor as a result of the numbers of machines in use during the latter part of the war. Also, they could see that the economics of using a tractor or a steam plough were heavily in favour of the tractor. For example, the fuel cost per acre when ploughing with a tractor was very much lower than for a steam ploughing set and the labour cost saving was obvious - one man against a 'gang' of five.

The need for the services of a steam ploughing contractor rapidly diminished and of the two firms mentioned earlier, Ward & Dale were wound up in 1939 and Smoothys sold its ploughing engines in 1940, although it continued steam rolling contracting until 1946.

There was one area of work for which steam ploughing engines, and their method of operation, were ideally suited, which did prolong the working life for some of them - dredging.

It is believed that the firm of Messrs. Mornement & Ray of East Harling, Norfolk, was first in this field, in 1899, and specialised in work on the tidal estuary of the river Ouse. Rivers, lakes, ponds and the Fens were also dredged and even today there are still one or two contractors carrying out steam plough dredging.

*Smoothy's Fowler ploughing engine Reg. No.HK9842 Fleet No.3.*

*Moving the Church at Leigh by Smoothy's traction engine and workmen.* Derrick Smoothy Collection.

TELEGRAMS: "SMOOTHY ROCHFORD, ESSEX." ESTABLISHED 1879 TELEPHONE NO. 5 ROCHFORD

From . .

# W. G. SMOOTHY & SON,

*Steam Plough & Engineering Works,*

**ROCHFORD, ESSEX.**

To

191

**Engineers.**

**Boiler-makers.**

**Contractors for Steam Ploughing.**

**Cultivating.**

**Harrowing.**

**Mole Draining.**

**Road Rolling and Scarifying.**

**Haulage undertaken by Steam Tractors and Traction Engine.**

**Engines on Hire, also Bought, Sold, or Exchanged.**

*William George Smoothy 1845-1908.*

# W. G. Smoothy

Although this part of the book is essentially about the activities of a Steam Engine Contractor it may not be out of place to start with a few words about an industry that has been practised for thousands of years - namely Tanning. This is a process whereby animal skins are turned into leather which can then be used for a wide variety of artifacts.

In the past this necessitated the extraction of tannin from vegetable matter, commonly the bark of the oak tree. After having been felled the bark was stripped from the tree. It was then shredded, ground up and mixed in a tank of fresh water and left until the tannin secretions were dissolved-out to the appropriate density required for the tanning operation. The liquid was then drawn off into large, clay-lined pits into which the animal skins were hung. Many pits were involved, allowing the skins to be gradually moved, over a period of several months, through solutions of increasing strength to ensure even tanning throughout the skins.

*Eddington & Steevenson traction engine Reg No. NO4495. Taken c1925 at Highbury Barn, Gt Cornard, Suffolk.* EATEC Archives

When this process had been completed, the resultant leather was scrubbed, dried then treated (curried) to increase its softness before being delivered to the leather worker.

Up until the middle of the last century many villages had their own tanneries but as the process involved was very slow and labour intensive, alternative methods of tanning took over and now there is only one traditional tannery operating in this country.

One of the centres of population that had its own tannery was Chelmsford, in Essex, close to the river Cam and this operated well into the 20th. century.

By now you, the reader, may be wondering how the business of tanning is relative to the theme of this section of the book and why a particular tannery has been mentioned.

The answer is twofold. Firstly because leather was widely used in connection with steam engines and enginemen and secondly, and more relevant, because of the involvement of a certain small boy in the tanning process.

Leather was used almost everywhere where steam engines, threshing machines and their operators were involved; to make belts, buskins, boots, aprons, gloves, sharpening strops, horse harness, reins, money pouches and the like.

As mentioned earlier, the tanning industry was very labour intensive which doubtless, a century or more ago, led to the employment of child labour.

One such boy was William George Smoothy born on May 12th. 1845 at Widford, near Chelmsford. He was one of a family of eleven children and at the tender age of nine he worked as a bark stripper at the Chelmsford Tannery of John Dixon & John Mathew Voss where his father William was a buyer of bark.

It is interesting to note that the tannery had a fixed steam engine and that, in 1858, together with thirteen other firms in the area, it was castigated by the local authorities as these engines created a smoke nuisance.

Two of those thirteen firms were Messrs. Eddington & Co. and Coleman & Morton and both were to have connections with William Smoothy.

It is not known how long William worked in the tannery but he did serve his apprenticeship with Messrs. Eddington (later to become Eddington & Steevenson) and then went to work for Coleman & Morton as a leading hand in their boiler shop.

It is highly probable that William acquired an interest in steam ploughing and steam rolling whilst working for these two firms as both were developing and producing associated equipment.

Eddingtons, besides producing a convertible steam roller, established the "windlass" method of ploughing which basically consisted of a two-tier wooden frame mounted on four cast wheels. On the upper tier a Portable steam engine could be placed which was then belt-coupled to a winding drum positioned on the lower tier and a rope from this could be fixed to a plough or mole drainer that was to be hauled across a field.

Coleman & Morton had also been involved with steam ploughing and it is recorded that on July 5th. 1856 they took part in trials at Writtle, near Chelmsford, at which one of their ploughs was drawn by a

Boydell Traction Engine. Incidentally, this engine steamed down from London on the day previous to the trials and completed the journey in nine hours. On completion of the trials the Boydell engine towed the plough back to Coleman's Yard and it was the first steam engine to be seen on the streets of Chelmsford.

From the foregoing it is evident that William had the opportunities to appreciate the ins and outs of the engineering industry and steam ploughing and haulage in particular, which must have stood him in good stead in the following years.

It is thought that William continued to work for Coleman & Morton until about 1870 as in a letter written by him from Blackfriars Place, Chelmsford, dated November 24th 1868 to his fiancé, Sarah Elizabeth Scaggs, of Rochford, Essex he bemoaned the fact that as he was a journeyman travelling widely around the Chelmsford area there were limited opportunities to see her. The letter also said that there did not appear to be any suitable jobs going in the Rochford area.

However, circumstances must have changed soon after he wrote that letter as he married Elizabeth in 1870 and about that time he was employed by the Rochford engineering firm of Messrs. Carey, Stilwell & Birch which, sometime after 1867, had erected a Steam Works on Lodge field.

He continued to work for Messrs. Carey, Stilwell & Birch for a few years thereafter but in 1879, at the age of 34, he thought his skills and knowledge were sufficient, and the time was ripe, to go into business on his own account.

At some stage he acquired premises in Weir Pond Road and in 1892 he bought, for £600, the King's Hill complex, bordering on East Street and Weir Pond Road which consisted of a large house and two semi-detached cottages.

Presumably, initially, William concentrated in his business on the repair of agricultural implements, but by 1892 he may have branched out into steam ploughing (if not also steam rolling) contracting. Evidence for this is supplied in a mortgage document, dated January 12th. 1893, between himself and William Peter Meeson of Doggets, Rochford.

This transaction was for £310.7s. of which £100 was to pay off a loan for the purchase of King's Hill and £210. 7s. which was outstanding on the purchase of Ploughing equipment. This is listed on the document as Steam Ploughing Tackle consisting of four Fowler Engines with ropes attached, two ploughs, two cultivators, one harrow, and two timber houses on wheels (living vans?) – these items were the security for the loan of £310. 7s.

*North Street, Rochford, looking south.*

John Smoothy Collection

*High Street, Rochford*

John Smoothy Collection

# W.G. Smoothy & Son - Personnel

William George Smoothy married Elizabeth Scaggs in 1870 and later that year they had a son, Harry William Scaggs Smoothy. Elizabeth died in 1871 and William re-married in 1873 to Emma Trupp. Harry was the sole offspring of the two marriages and in 1879, although only a child, he was made a partner with his father when the firm of W.G. Smoothy & Son was formed. He was therefore, at that early age, destined to take over the business, which he did in 1908 when his father died and he successfully ran it until 1946.

Harry founded the Steam Cultivation Development Association in London in 1911, which proved of great value to the country in the 1914-18 war by its knowledge of steam utilisation for driving vehicles through the mud on the Western Front. He was the Essex representative of the National Road Roller Association and a member of the National Traction Engine Users Association.

He was elected onto the Rochford Parish Council in 1896 and celebrated his 50th. year of membership in 1946. Also, in 1931, he was elected a member of the Rochford District Council and retired due to ill health a few months before passing away on September 7th. 1949. In his younger days he was a champion cyclist and won many races in the Rochford area and he had the distinction of being the first owner of a pneumatic-tyred bicycle in Essex. Bowling was another of his interests.

*Rochford Station. This train is believed to be the first G.E.R. engine to run between London and Southend c1895. Members of the Smoothy family are on the platform.*
John Smoothy Collection

On the November 26th. 1892 he had married Ada Annie Wadey at Edmonton, London, and the couple eventually had five children. Oswald (in 1892), Norman (in 1894), Cyril (in 1899),

Frank (in 1902) and Raymond (in 1908). All, except Raymond, contributed in some measure to the success of the firm.

Oswald became an accountant and went into partnership with his father, Norman and Frank worked mostly as steam roller and steam plough drivers, and Cyril became the firm's carpenter and joiner. Raymond enlisted at an early age in the Royal Air Force and after serving his time he worked for a very short while in the firm's office before re-joining the Royal Air Force during the second world war. Cyril left the firm in 1936 and went to work for the Rochford District Council at their Sewage Works. At some stage early in the 1940's, Frank and his wife formed the Stambridge Threshing Company which lasted until Frank died on October 25th. 1963. Incidentally, when the effects of the firm were auctioned off on May 27th. 1964, Jim White, who once worked for Smoothys, bought a "Contractor's Living Van" for £52. 10s. This van had been made by Smoothys.

Norman Smoothy (father of John, mentioned in the acknowledgements) joined the Royal Engineers soon after the outbreak of World War II and drove a steam roller; he refused promotion so that he could continue driving rollers. During his service career he rolled runway extensions at Reykjavik, Iceland, and in Gibraltar and also did a lot of rolling work with the British Expeditionary Force in Europe.

Whilst with the firm Norman fitted extra wide rolls to their two 2 1/2 ton Aveling & Barford petrol rollers so that they could be used on local grass tennis courts. Apart from the immediate family Smoothys employed, over a period, many steam roller and ploughing engine drivers, several of whom came from the East Essex Steam Roller Association, and one of these was Jim White, mentioned above, a nonagenarian now living at Great Leighs near Chelmsford.

Jim was Smoothy's youngest driver and although it is over 60 years since he left the firm his active and retentive memory can recall much information about the firm as follows:

Mr. Smoothy was a stickler for having his engines kept in beautiful working order and would not tolerate any driver who did not look after his machine. Whenever he bought a second hand roller the first thing he did was to take off the water pump and replace it with a low pressure injector. As well as the high and low pressure injectors all of Smoothy's rollers were equipped with a Morrison Scarifier, oil pump, water sprayer, awning, water lift and hose, loose tools and cover, a 10 ton bottle jack, a wash-out pump, 15 ft. chain, three roller lamps and warning plates. The Living Vans, a number of which were built by Smoothys (mainly by

LAUNCH OF THE ROYAL SOVEREIGN

*The Royal Sovereign, an 80 ton vessel fitted with twin screw six cylinder engines each of 70hp, capable of carrying 250 people, was launched at Southend on May 27th 1924. She was built in the yard of Messrs Hayward & Croxson and, on May 24th, was towed along the esplanade by one of Smoothy's largest ploughing engines Reg. No. HK9839, Fleet No.8. It was driven by Norman Smoothy to a position on the beach where the Royal Sovereign could be launched. The engine was then placed behind the ship and its rope was run out to a pulley anchored well down the water line, then back up to the Royal Sovereign.*

*Initially the anchorage broke when winding-in started and after a couple of days a two-pulley system was successfully used. As a local paper put it "With beautiful grace and dignity the ship glided down as if she was built on a slipway, breaking the water in fine style". A novel use for a ploughing engine?*

John Smoothy Collection.

Telegrams:—"SMOOTHY, ROCHFORD, ESSEX."

56151 ROCHFORD
~~TELEPHONE 51 ROCHFORD~~
~~Telephone No. 5 Rochford~~

# W. G. SMOOTHY & SON,

## Engineers, Steam Plough, Roller and Traction Engine Proprietors,

### ROCHFORD, ESSEX.

# INSTRUCTIONS TO DRIVERS.

1. The Engine you are in charge of is well found with all tools, lamps, notice boards, etc., and other necessaries for proper and safe working, and must be kept thoroughly clean and in good order, washed out regularly every fortnight, and all small repairs done to keep the Engine in proper working order without losing time.

2. Any defective plant, or in case of accident or breakdown or other repairs necessary otherwise than what you can immediately rectify yourself, notify at once to the Works, or if urgent ~~telephone or~~ wire:—

**"SMOOTHY," ROCHFORD, ESSEX,**

**stating where and when to send boilermaker or fitter, and flagman must meet trains when possible.**

3. **ACCIDENTS.**—Should an accident occur or any damage be done to another person's property, whether through the driver's fault or not, the driver must at once obtain the names and addresses of any witnesses present, get a policeman's evidence and take his number, so that any claim for damages can be resisted or made by the Firm, and the case must be reported at once to the Works, especially in the case of personal injuries.

4. Drivers will render themselves liable to instant dismissal with loss of any wages that may be due to them for any of the following offences:—Using an iron plug, white or red lead or anything other than proper fusible metal, in safety plug of fire box. Reducing ferrules or otherwise tampering with safety valves, and for being intoxicated during working hours.

5. **FUEL.**—As we have contracts for supply, ask for instructions where to order. It is particularly requested that drivers and flagmen should use every care to prevent waste, and excessive smoke.

6. Roller Drivers must always use Coke when working in Towns and populous places.

7. Notice Boards must be exhibited at least 100 yards distant from each end of the Roller, when at work.

8. Drivers leaving will receive or must give a full week's notice to expire Saturdays.

9. Flagmen are not allowed to drive at any time without special permission.

10. When working for others than Councils, drivers must get their time ticket signed by a foreman or person in charge and send to Works each week with time sheet.

11. The working hours are recognised as the same as roadmen in each district.

12. It is to be understood that drivers and other employees in accepting these instructions will pledge themselves to a proper observance of them, and when working in Yard or Works, will be subject to shop rules and time, &c.

*Read carefully and keep for reference these instructions, also the Laws relating to Road Engines, and special attention is called to Clause 21.*

Cyril Smoothy) were equipped with a nice coal-burning stove, an oven and cooking utensils, coal and oil lockers, mattresses, pillows and blankets.

Jim reckoned that the Smoothy firm had the best maintained fleet of rollers anywhere in Essex.

In his own words Jim said "When the Governor (Harry) was dressed up to do business with the various Councils he really looked like a country squire, with his nanny-goat beard, plus fours and thick woollen socks turned over below the knee. I never did see him drive an engine, nor even get into the driving position".

Jim started with the firm in the mid 1930's and his first job was driving a 10 ton Greens roller Reg. No. NO 4613 Works No. 1420 which he said had the cylinder on the flywheel side as opposed to the normal type with it on the right hand, or gear, side.

He, along with the other drivers, was responsible for ordering his own coal and coke and Smoothys expected that he would burn a half ton of each every fortnight. This meant careful juggling with the proportions used and Jim said he kept a good 12" thick fire going. He used the coal only to steam up in the mornings and to bank the fire up at night plus a little to put on the fire when he stopped for lunch.

*Aveling & Porter steam roller. Taken whilst working for East Suffolk CC.*
Derrick Smoothy Collection

The area covered by Smoothy's contracting work, a lot of which was for local authorities, ranged towards London up to Romford and northwards as far as the Suffolk border - a very large area. Jim was lucky in a sense as most of his work was on behalf of the Chelmsford Rural District Council, which meant that he was close to home and did not need to use a living van for sleeping purposes. He said that the busiest jobs that a roller driver had to contend with was working on water-bound roads (described elsewhere) and Smoothys expected that a roller "gang" would slurry-in 100 tons of granite chippings a week, with the average job taking 300 tons to complete. This granite rolling lasted until the end of March, then, when the weather turned warmer, tarring and shingling would start.

For a short period Jim drove a Smoothy 6 ton petrol/paraffin Aveling & Barford tandem roller, working behind a tarpot. This machine had a tin roof and during a hot summer's day sitting under this awning close to the radiator, Jim said it was sometimes difficult to keep awake and he was glad to get back onto steam engines.

For a while Jim was contracted out by Smoothys to work for Crawford & Co. Public Works Contractors, rolling sewer trenches. He said that Crawfords had two steam rollers of their own; one was a 12 or 14 ton Aveling & Porter with the Fowler compound pattern cylinders and the other was a single crank compound Burrell. Both were is a shocking state, the Burrell being especially bad, blowing out steam between the cylinder seating and the top of the boiler barrel. Soon after Jim finished this contract Smoothys claimed both of these engines in lieu of a debt. They were then stripped down completely, thoroughly overhauled and painted and made as good as new.

Jim recalls the following people as some of the firm's drivers during his time with Smoothys up to the summer of 1940 when, due to the illness of his father, he left and re-joined his family's threshing business:

| | |
|---|---|
| Ben Newall | - Roller driver, mainly working on Romford Borough Council contracts. |
| Fred Holmes | - Roller driver, mainly of Fleet No. 9 'Monarch' an A & P roller Reg. No. NO 3270. Fred lived at Purley. |
| Fred Farreno | - Driver of Fleet No. 26, Fowler roller Reg. No. NY 2227. |
| Jack Copping | - Driver of SCC Burrell roller. He lives at Rettendon. |
| Georgie Brown | - Roller driver, mainly of Green's engines. He lived at Rochford. |
| Bill Harvey | - Roller and steam plough driver. |
| Bill Everett | - Roller driver |
| Dick Mansthorpe | - Foreman of a steam plough gang. |
| Fred Sewell | - Foreman of a steam plough gang. |

In addition to the above Jim recalls that George Harris was employed as a Blacksmith and some of the work involved drawing-out and repointing scarifier tines and making new tyres for vans, water carts and the like. Harold Fisk was a full time fitter and turner, maintaining the engines and allied equipment and Joe Scroggins was a boiler maker.

Dennis Letch, of Capel St. Mary, Essex, recalls that his grandfather, Charles Letch, went to work for Smoothys at the age of 21, in 1926. After a few weeks working in the engineering shop he was made up to a roller driver and mainly drove Fleet No. 9 Reg. No. NO 3270. Charlie also spent several years operating with a pair of Smoothy's ploughing engines. He then returned to steam rolling and was on hire from Smoothys to Leighton Contractors Ltd. of Billericay, Essex, driving a roller at the Burstead New Sewage Works. He left Smoothys in 1936.

Another driver of N0 3270 was Herbert (Bert) Lewis Smith, born in 1899, who joined Smoothys soon after being released from a World War I Prisoner of War Camp. According to his son, George, his father worked for Smoothys over most of Essex and one of his jobs involved taking his steam roller onto Canvey Island at low tide, as at that time there was no bridge.

Sometime after the end of World War II, Bert set himself up as a Steam Roller Contractor and in 1946 he attended the Smoothy auction hoping to buy the Aveling & Porter roller NO 3270, but he was outbid when it went for £95!

*Aveling & Porter steam roller Reg No. NO3270 Fleet No.9. Charles Letch is the driver.*

Photo D.C. Letch Collection.

*Thomas Green & Sons Fleet No.27. Taken on July 6th 1915 whilst working at Much Hadham for the Herts CC.*

Derrick Smoothy Collection

*Aveling & Porter steam roller Fleet No.18. Norman Smoothy is on the extreme right. Taken on Sept 29th 1913 whilst working at Stowupland Road, Stowmarket for East Suffolk CC.*

Derrick Smoothy Collection

# Smoothy's Engines

A firm's Engine Register can be a veritable gold mine of facts and figures about individual machines, which are priceless to an interested archivist. However, a drawback to a Register is that it may not be complete, may contain mistakes or may not be comprehensive.

Smoothy's Engine Registers, of which there are three covering Steam Rollers, Motor Rollers and Steam Ploughs, do have the drawback of not being complete.

A case in point is the record sheet of the engine listed as Fleet No.25, an 1883 Aveling & Porter Steam roller which had been delivered to Smoothys in 1910. A note against it reads "Rec'd. here by road. S. Lawler driver July 22nd. 1910. Bought at Henry Lewis & Sons yd. Reading (No 23 steam Wagon delivered in exchange)" There is no mention of a Steam Wagon in Smoothy's records but there is a Fleet No.23 on their roller register and this refers to a machine bought in 1926. However, a photograph of a Mann steam wagon, with W.G. Smoothy & Son on it, does carry the number 23.

*A Mann steam wagon Fleet No.23. This was delivered to Henry Lewis & Sons, Reading, in exchange for steam roller Fleet No.25 (159 Berkshire).* John Smoothy Collection.

There are other omissions as it is known that the firm did have some traction engines and a Super Sentinel. It also appears to have been Smoothys practice to use identical Fleet numbers for different machines, but seemingly only after the earlier engine had been disposed of. Examples are, with the two

Smoothy purchase dates, Fleet No.14 (1905 and 1932), Fleet No.18 (1906 and 1934), Fleet No.25 (1910 and 1933), Fleet No.28 (1913 and 1931). Also there are gaps in the sequence of Fleet numbers. Smoothy's Engine Register, with the page sizes of approximately 330 mm by 280 mm, are nevertheless quite comprehensive with information listed under up to 36 headings ranging from Fleet numbers to such items as Driving Pins, Gauge Glasses, Boiler Tubes, Mud Holes and the like. Exact measurements and other relevant details are given, such as when replacement items were fitted. As an example, on the record sheet for Fowler ploughing engines Nos. 11636/7 (Fleet Nos. 3 & 4) against the heading Drum is the following: "(90 teeth) 11636 800 yards ex Hull engine 12240. 11637 500 yards 3/4" diameter 30 wires, takes 10 rounds. $9^{3}/_{4}$" deep $7^{3}/_{4}$" width (flange to flange)"

Of interest there is a Quotation, c1916 from Fowler for spares for Ploughing engines 3379 / 80 as follows: "Steel Road Spur Pinion £5 - 0s. - 0d. 3rd. Motion wheel cast iron £6. 2s. 6d. in steel £23. 10s. 0d. 3rd. Motion Shaft £6. 16s. 0d. For engine 13481 (Belonging to Keeling) a Chimney barrel £2. 17s. 6d. and for engines 1404 / 5 (there is an error on the sheet as 1404, is a Traction engine number - it should be 1402), Front Axle, square £7. 14s. 0d., Slide Valve £1. 6s. 6d."

Smoothys also had a few motor rollers but details of these are sketchy. Four were auctioned on May 23rd. 1946 as follows: Fleet No.38 Aveling & Porter 6 ton Diesel Works No. 12923 (sold for £775), Fleet No.36 Aveling & Barford $2^{1}/_{2}$ ton petrol engined (sold for £480), Fleet No.30 Barford & Perkins 5 ton (sold for £95) and Fleet

*Lined up for the auction of May 23rd 1946 in Smoothy's yard.*

The John Mullet Collection

*Wallis & Steevens Fleet No.22. Harry Smoothy is on the right.*

Derrick Smoothy Collection

*Aveling & Porter steam roller Fleet No.25 with Norman Smoothy at the controls and Cyril Smoothy standing. Taken c1915.*

Derrick Smoothy Collection

No.12 Barford & Perkins Reg. No. EV 5162 2½ ton petrol engined (sold for £330).

The engine lists in this chapter have been compiled from Smoothys own Engine Register and from RLS details compiled by the late Alan Duke. These two lists may not be comprehensive, especially as they show that the earliest Ploughing engine was obtained in 1906. As mentioned elsewhere, a mortgage document, dated 12/1/1893 states that W.G. Smoothy has put up as security four Fowler Ploughing engines, plus other tackle.

Alan Duke also kept a separate list of engines that could not be identified by a Makers number, and one of these was an engine Reg. No. NO 6644 which by August 1922 had been registered by W.G. Smoothy & Son, of Rochford and was last licensed in 1924. Alan was unable to find out if this was a Ploughing engine or a Traction engine.

The extent of the firms operations in the Steam Roller Contracting business can be gauged by the number of engines and associated equipment that were auctioned off on May 23rd. 1946 by Henry Butcher & Co. There were 16 steam rollers (almost all with 10 ton bottle-jacks, a wash-out pump, a 12 or 15 feet chain sling and three roller lamps), 18 Living Vans, 22 Water Carts, mostly Baker types, 4 Motor rollers, a Burrell traction engine Works No.2252 and a Clayton & Shuttleworth 4 nhp Portable.

Incidentally, the end of the firm did not come about due to a lack of business but because Harry Smoothy wished to retire, as he was 75 years old, and his five sons could not agree as to how the business should be run. Harry therefore made the decision to close down the firm.

*Aveling & Porter Reg No. HJ5751. It was Smoothy's Fleet No.34 and it is shown being restored in Jack Shuttlewood's yard, near Braintree, Essex.*

Photo: Roy Daniels

*Lined up for the auction of May 23rd, 1946, in Smoothy's yard. From the left Burrell KE4297 Fleet No.15, Aveling & Porter NO3605 Fleet No.17, Aveling & Porter XC9522 Fleet No.25. All sold for £100 each.*

The John Mullett Collection.

*Aveling & Porter steam rollers Reg. No. NO3326 Fleet No.29 and Reg. No. HJ1732 Fleet No.23 at Smoothy's auction May 23rd 1946.*

The John Mullet Collection.

# LIST OF STEAM ROAD ROLLERS

| Fleet | Date Bought | Reg. No. | Engine No. | Manufacturer | Type | Weight | Year Made | Sale Price |
|---|---|---|---|---|---|---|---|---|
| 1 | 29.7.1909 | 9 ECC | 2982 | Aveling & Porter | SC | | 1892 | |
| 3-10 | 4.10.02 new | NO 3271 | 5013 | Aveling & Porter | SC | 10 ton | 1902 | £100 |
| 5 | 1.12.1902 | 308 ECC | 1500 | Thos. Green & Son | SC | | 1891 | |
| 7 | 17.2.1903 | 312 ECC | 2221 | Aveling & Porter | SC | | 1886 | |
| 9 | 4.1.04 new | NO 3270 | 5423 | Aveling & Porter | SC | 10 ton | 1904 | £95 |
| 11 | 31.1.1922 | NO 4613 | 1420 | Thos. Green & Son | | 10 ton | 1895 | |
| 12 | 25.5.1904 | 35 Middlesex | 1735 | Aveling & Porter | SC | | 1882 | |
| 14 | 20.2.1905 | 359 Essex CC | 4261 | Aveling & Porter | CC | 10 ton | 1899 | £110 |
| 14 | 14.5.1932 | ET 1952 | 4750 | Aveling & Porter | | 10 ton | 1901 | £110 |
| 15 | July 1925 | KE 4297 | 2702 | Chas.Burrell & Son | SC | 9/10 ton | 1904 | £100 |
| 16 | 23.12.1905 | 377 Essex CC | 2485 | Aveling & Porter | CC | | 1889 | |
| 17 | 20.10.1923 | NO 3605 | 3195 | Aveling & Porter | SC | 12 ton | 1893 | £100 |
| 18 | 4.4.1906 | 14 London CC | 2438 | Aveling & Porter | CC | 7/8 ton | 1888 | |
| 18 | 19.12.1934 | KT 3843 | 8447 | Aveling & Porter | | | 1914 | £135 |
| 19 | 22.11.1927 | U 1235 | 12107 | John Fowler & Co. | | 10 ton | 1910 | |
| 20 | June1924 | DL 2580 | 2260 | Wallis & Steevens | SC | 10 ton | 1894 | £100 |
| 21 | 10.3.1907 | NO 3272 | 2106 | Aveling & Porter | SC | 9/10 ton | 1885 | £95 |
| 22 | 13.6.08 new | 455 Essex CC | 2986 | Wallis & Steevens | CC | | 1908 | |
| 23 | 28.4.1926 | HJ 1732 | 8672 | Aveling & Porter | SC | 10 ton | 1915 | £95 |
| 24 | 6.1.1909 | 456 ECC | 3106 | Aveling & Porter | SC | | 1892 | |
| 24 | c.1930 | E 5352 | 2328 | Chas. Burrell & Son | SC | 10 ton | 1900 | £120 |
| 25 | 22.7.1910 | 159 Berkshire | 1935 | Aveling & Porter | SC | 9/10 ton | 1883 | |
| 25 | June 1933 | XC 9522 | 3522 | Aveling & Porter | SC | 10 ton | 1895 | £100 |
| 26 | 16.1.1928 | NY 2227 | 10958 | John Fowler & Co. | CC | 10 ton | 1907 | £100 |
| 27 | 10.1.1912 | 540 ECC | 1399 | Thomas Green & Son | SC | | 1899 | |
| 27 | 20.10.1930 | NO 1492 | 3478 | Aveling & Porter | | | 1894 | |
| 28 | 18.6.1913 | 374 Essex | 1648 | Aveling & Porter | SC | | 1880 | |
| 28 | Feb. 1931 | NO 3253 | 40102 | Clayton & Shuttleworth | CC | 10 ton | 1908 | £105 |
| 29 | 6.7.1914 | NO 3326 | 5853 | Aveling & Porter | CC | 8 ton | 1905 | £260 |
| 32 | 20.2.1937 | XD 8048 | 7284 | Wallis & Steevens | SC | 8 ton | 1912 | £150 |
| 34 | June 1942 | HJ 5751 | 11527 | Aveling & Porter | CC | 8 ton | 1926 | £550 |

# LIST OF PLOUGHING ENGINES

| Fleet No. | Date Bought | Reg. No. | Engine No. | Manufacturer | Type | H.P. | Year Made |
|---|---|---|---|---|---|---|---|
| No. 2 { | 1906 | NO 4920 | 3379 | J. Fowler & Co. | Single | 14 | 1878 |
| Set | 1906 | NO 4921 | 3380 | J. Fowler & Co. | Single | 14 | 1878 |
| 6+ | c.1915 | 369 ECC | 1402 = | J. Fowler & Co. | Single | 14 | 1870 |
| 8+ | c.1915 | 370 ECC | 1405 = | J. Fowler & Co. | Single | 14 | 1870 |
| 6 | 31/12/1917 | HK 9838 | 15242 * | J. Fowler & Co. | AA7 compound | 16 | 1917 |
| 8 | 31/12/1917 | HK 9839 | 15243 * | J. Fowler & Co. | AA7 compound | 16 | 1917 |
| 5 | c.1919 | HK 9840 | 14213 * | J. Fowler & Co. | BB compound | 14 | 1914 |
| 7 | c.1919 | HK 9841 | 14214 * | J. Fowler & Co. | BB compound | 14 | 1914 |
| 3 | 1921 | HK 9842 | 11636 * | J. Fowler & Co. | B4 compound | 14 | 1909 |
| 4 | 1921 | HK 9843 | 11637 * | J. Fowler & Co. | B4 compound | 14 | 1909 |

* Sold at auction 10/7/1940

= Early serial numbers did not always follow in pairs.

\+ RLS records show nothing after c.1915 to W.G. Smoothy

Although Smoothy's records show they still held this pair in January 1917, they could have been scrapped or sold soon after and 15242/3 bought new – hence 6 & 8 re-used.

Refer to chapter on engines as this seems not to be a complete list.

# Smoothy's Workshop

In 1892 W.G. Smoothy bought, for £600, King's Hill, King's Hill Cottages and land to the west with frontage onto Weir Pond Road.

It was on this land to the west that he built his workshops as shown on the plan, which is not to scale.

The blacksmith's, general workshop and the machine shop were contained within a building about 150 feet long and 60 feet wide which had a pointed roof from which louvred vents protruded skywards. On the end of this building was the carpenter's shop which had its roof sloping down along the line of the above mentioned workshop, from about 20 feet down to seven feet above ground level. The only source of daylight for this shop was through skylights in this roof.

The 'Top Shed' was about 70 feet long by 50 feet wide and was constructed, as was the main workshop area, of timber faced with galvanised corrugated iron and was painted with red oxide. It was high enough for the easy entry of steam engines it had louvred vents in the apex of the roof to allow smoke and steam to exit. Within this building was storage space running along its entire length and was about 10 feet wide. The smaller store was used for storing oil, and the rest was the "Electrical Workshop". It was in this area that accumulator and battery charging took place.

Adjacent to the 'Top Shed' was an open area for the storage of coal and scrap iron.

Just outside the machine shop was a lean-to under which

*John Smoothy, in 1948, stands in front of the outside toilet of No.2 King's Hill Cottages. Behind him can be seen the sloping roof of the Carpenter's shop.*

John Smoothy Collection

there was was a Portable steam engine that drove the overhead shafting in the machine shop. By 1931 this source of driving power had been replaced by a large Petter double flywheel diesel engine housed within the workshop. The Portable remained for many years as a stand-by.

An office, garage and a lean-to were also within the Works complex.

Across Weir Pond Road from the Works was a piece of land that was acquired by Smoothys and about two thirds of this was given a hard surface and the other part was left grassed over. This area was known as "The Meadow" on which steam rollers, living vans and mobile metal water butts were kept.

*Marshall portable. It may have been used to drive the overhead shafting of Smoothy's workshop.*

Derrick Smoothy Collection

The Workshops were of sufficient size, and contained the necessary facilities and tools, to enable almost all the requirements to be met for the maintenance, repair and re-build needed to keep a fleet of steam engines in tiptop condition. New boilers, fire boxes and the like could not be made, but existing ones could be patched to a high standard and parts such as chimneys

*Part of Smoothy's Workshop. The door leads into the Carpenter's shop. The overhead shafting was originally driven by a Marshall portable engine, kept just outside the right of the building under a lean-to.*

Derrick Smoothy Collection

and barrels for water butts could be rolled. Living vans were made from scratch although their metal wheels had to be purchased.

There was a constant flow through the blacksmiths shop of scarifier and cultivator tines for sharpening and shaping.

Not only were the maintenance and repair needs of Smoothy's engines catered for in these workshops, but firms such as Keelings of Wickford and Bentalls of Southchurch also brought their engines there when work was needed on them.

At the present time there are new buildings on the old Smoothy site, which are owned by Ernest Doe & Sons Ltd. On one of the walls of their workshop is displayed a piece of a roof-beam and a couple of bolts with a plaque which says "This is the last reminder of our old workshop on this site".

*Aveling & Porter steam roller Fleet No.10 at Oldchurch Road, Romford, July 1936.*

Derrick Smoothy Collection

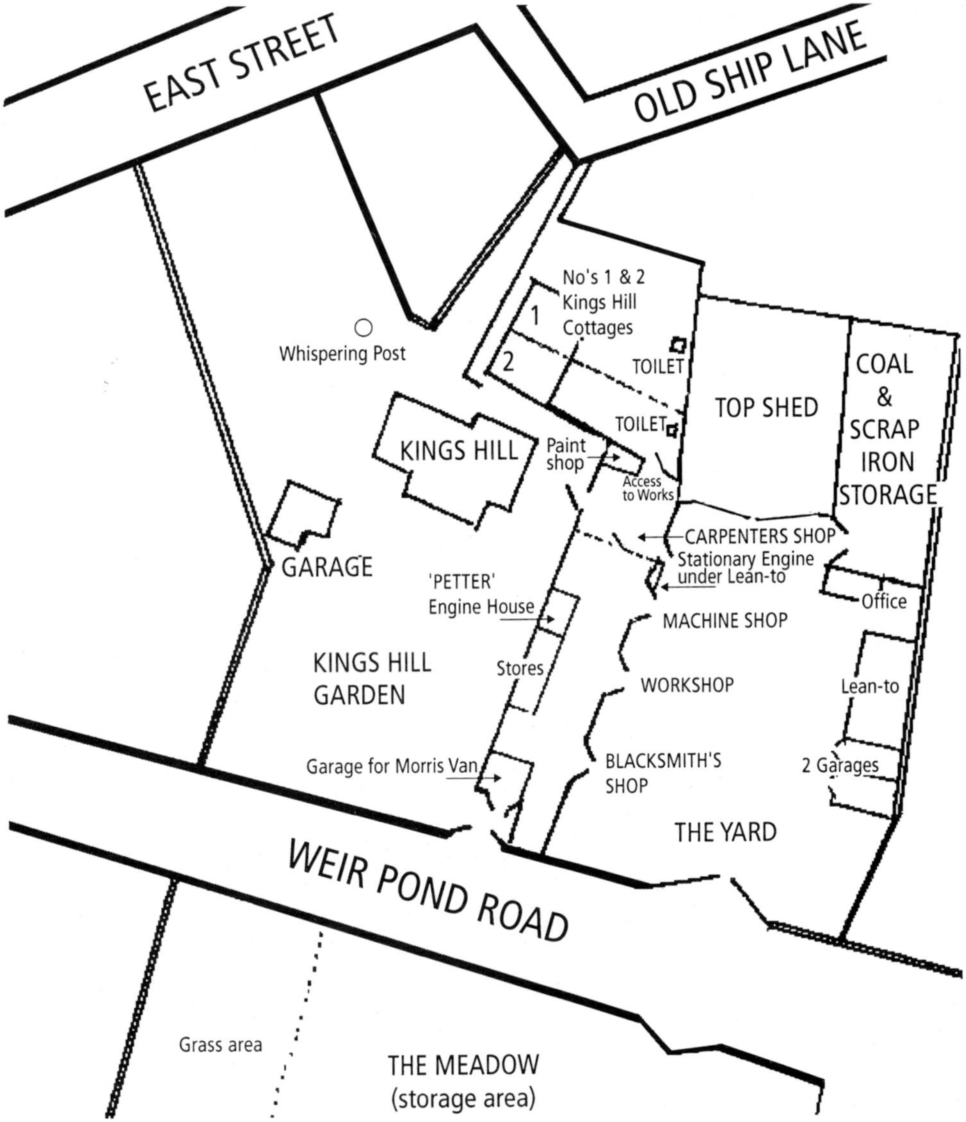
EAST STREET
OLD SHIP LANE
No's 1 & 2
Kings Hill
Cottages
1
2
Whispering Post
TOILET
TOILET
TOP SHED
COAL
&
SCRAP
IRON
STORAGE
KINGS HILL
Paint
shop
Access
to Works
CARPENTERS SHOP
Stationary Engine
under Lean-to
GARAGE
'PETTER'
Engine House
Office
MACHINE SHOP
KINGS HILL
GARDEN
Stores
WORKSHOP
Lean-to
Garage for Morris Van
BLACKSMITH'S
SHOP
2 Garages
THE YARD
WEIR POND ROAD
Grass area
THE MEADOW
(storage area)

# King's Hill and the Whispering Post

King's Hill, in East Street, is one of Rochford's most beautiful and historic houses, built in the 14th. century and owned by W.G. Smoothy from 1892 until 1939. When bought by William Smoothy, for £600, the Deeds stipulated "Subject to the right (if any) of the Lawless Court to assemble yearly round the Whispering Post (position shown on the plan)".

Tradition has it that the Lawless Court, otherwise known as the Whispering, King's or Cockcrowing Court, originated early in the 12th. century when a Lord of the Manor in Rayleigh, Essex, returned home after a long absence. He retired to bed but was awakened around midnight by the crowing of a cockerel and he then heard loud whispers from below his bedroom window.

He ascertained that some of his tenants, who seemingly had a grudge against him, were plotting to kill him, so he confronted them and as a punishment ordered them to give formal acknowledgement of feudal allegiance to him and to do this annually on each Wednesday after Michaelmas. They were to assemble around a post and at the first crowing of the cockerel at dawn they had to pay homage in whispers to the Lord of the Manor.

This post, just under two metres high and ten centimetres square, with its top shaped like a lighted candle, was inevitably called the Whispering Post. This court was held for several centuries at King's Hill, Rayleigh, and was then transferred by the 2nd. Earl of Warwick to a site at Rochford; and the Whispering Post was placed in front of a house there which was then named King's Hill.

At some stage over the years the Lord of the Manor provided, in a local hostelry, a substantial supper preceding the assembly of the Court. Later on, the Whispering Court turned into a magnificent event involving locals and onlookers from far and wide.

It would appear that the last Court was held in 1892 and all that is left as a reminder of an interesting event of long ago is the Whispering Post.

This has been moved at least twice from its original position at King's Hill, Rochford. Firstly it was in the front of the house, then it was moved around 1911/12 to the centre of a small flower bed at the side of the building, then to facilitate the entry of cars to the garage it was moved, sometime between 1926 and 1931, to its present position. A new post was made in 1934 by Cyril Smoothy as the original was rotting away.

In 1987 a plaque was erected by the Rochford Historical Society, on the side-wall of King's Hill, which commemorates the assembly of the Lawless Court around the Whispering Post from circa 1630 to circa 1892.

In 1967 King's Hill was purchased by Rochford District Council for £25,000; they wanted to sell it as a building site but encountered a lot of local opposition. In 1985 they sold it for £129,950 and it has since remained a private house.

*King's Hill showing the Whispering Post in its original position. Inset – a later picture which shows the new position of the post and alterations to the front of the house.* John Smoothy Collection

*Smoothy's roller and road making gang at Rawreth on the A130.* Photo: Brenda Hurren Collection.

*King's Hill. The Whispering Post can be seen through the tree. The plaque is to the right of the door.*
Photo: Pat Freeman

*St. Mary's Church, Debenham.*
Photo: Pat Freeman

*Aveling & Porter Tandem steam roller W.No.7411. This is the only survivor from the nine of this type built. Taken in the yard of Mick Smith, Hurst.*

Photo: David Spooner

*Garrett Showmans engine 'Lord George' W.No.33358. It was built as a 4CD tractor but was converted, in 1921, by Bloomfields and went to F. Harris, a Showman of Clare. The late George Swann bought it in 1954 and it is now owned by his son Kevin.*

Photo: Kevin Swann Collection.

*Aveling & Porter steam roller "Monarch" ex-Smoothy's Fleet No.23 with Jim White ex-Smoothy's steam roller driver. Taken at Barleyland's Steam Rally in 1998.*

Photo: Pat Freeman

*"Bunty" Garrett No.34641 once owned by Bloomfields, now at Bressingham Steam Museum, with its regular driver Maurice Ruffles.*

Photo: Bressingham Museum

*The three Bloomfield brothers, Neil, Russell and Keith.*

Photo: Pat Freeman

*Keith Bloomfield on one of the four seats that he designed and made for Debenham Parish Council. "Old Tiles" is to the left.*

Photo: Pat Freeman

*Garrett W.No33295 owned by J.W. & A Pearl of Otley, ex H.Thurston & Sons (which explains the twisted brasses), hauling a tree trunk out of a pond on April 19th 1935.*
Photo: Kevin Swann Collection.

*Sentinel Roadless Tractor. Agents William Jacks & Co. This photo was found in the Smoothy's records, but no other details are available.*
Derrick Smoothy Collection.

*Interested onlookers at a demonstration of Bloomfield's mole pipe-layer modification.*

Telegrams: "Bloomfield, Debenham" Telephone: Debenham 53

**M. Bloomfield,** *General Engineer*

ASPALL STATION, L.N.E.R. DEBENHAM, STOWMARKET

*Malcolm Bloomfield 1891-1976*

# M. Bloomfield

Malcolm Bloomfield was born at Ashfield, Suffolk, in 1891, the son of Stephen Bloomfield, farmer and coachbuilder and Victoria (née) Bennett and it would be true to say that as a boy he was more interested in things mechanical than farming or, indeed, what could be learned at school.

An early story relates that the blacksmith at Ashfield, Fred Fairweather, took the young Malcolm on the cross-bar of his bicycle to a nearby funfair where Malcolm saw his first real Showmans Traction engine. The sight of this magnificent machine generated in him a permanent interest in Steam and, on leaving school, his father wisely took his son to see Sir Frank Garrett at the Leiston Works where an apprenticeship as an engine turner and fitter was agreed. Mr. Bloomfield deposited the sum of £10.00 as a guarantee of his son's good conduct for the apprenticeship period which was recorded on his Indenture to run for six and a half years from March 20th. 1906 to September 20th. 1912. The Indenture was signed by Malcolm Bloomfield, Stephen Bloomfield, Frank Garrett and Alfred Garrett.

*Entrance to Debenham High Street. The Forrester's Hall is on the left and Bloomfield's Cottages are on the right.*
Neil Bloomfield Collection

Whilst working there he lodged with a family in Leiston and at weekends he cycled home to the family farm at Ashfield-cum-Thorpe. He became firm friends with a Stanley Cooper, a neighbour and watchmaker of about the same age, and this friendship endured for over fifty years! In their spare time Malcolm and Stanley somehow contrived to fit an engine and propeller shaft into a standard rowing boat. By all accounts this was quite successful and was hailed as a "first" for these two young men.

Malcolm completed his apprenticeship in 1912 and left Richard Garrett & Sons with savings of £14.00. Whether this sum included the refund of the £10.00 bond is not known but he considered this sufficient to consider starting his own business.

He had his eye on a little-used forge in Debenham at the junction of High Street and Kenton Road, which later was to be called 'Bloomfields Corner', but initially had to settle for some space a little further down the High Street with a millwright by the name of Sam Clarke. There he learned a lot from Sam who no doubt gave this young man sympathetic encouragement. In due course Malcolm was able to rent the forge that he had his eye on and really got down to business, serving the local needs in a small way.

*To the right of the Public House is the old baker's shop and house and the first of Bloowmfield's cottages can be seen.* Picture: Pat Freeman.

In 1913 he married Valentine May Abbott, daughter of Mr. Abbott of Monk Soham, a local carrier and shopkeeper and they set up house in one of the cottages on the corner site.

Malcolm by now was a busy young man, driving the forge and a small lathe with an old motor bike engine. His skills were such that he became a valuable asset to the local community, repairing farm equipment such as steam mills and engines and turning his hand to repair smaller things such as clocks and watches. In this respect he was given "Exempt Occupation" status during the 1914-1918 war.

The corner site was wholly owned by a Mr. Balls who lived there and ran a small bakery at one end but in 1919 he apparently defaulted on his mortgage and moved out and Malcolm bought the whole site for £350.00. This comprised of the forge and buildings with the yard, two cottages, the bakery and house.

Mrs. Bloomfield kept the bakery going for a time, selling bread in the little shop for one farthing profit per loaf, and other items such as toys, watches and everyday necessities such as gramophone needles and the like.

At the same time she looked after three young children (although with some domestic help) and also took care of the book-keeping side of the business. It is interesting to note that, in keeping with country tradition, the family kept a couple of pigs in a small sty up until about 1935.

*This Burrell is one of the first engines to enter Bloomfield's yard. Malcolm Bloomfield is on the engine. Ted Abbot, his brother-in-law, is at the front.* Neil Bloomfield Collection.

Meanwhile, Malcolm had established a viable and busy concern, using his skills to good effect in servicing and repairing portable and traction engines, threshing machines, wagons, tumbrils and other farm implements and tools. He was particularly proud to fully overhaul, and repaint in their original livery, traction engines and he also began to repair engines and equipment used by travelling showmen.

In 1920 a mains gas supply was laid in the village from a small gas works at the junction of Water Lane and Priory Lane (then known as Hog's Close Lane). A gas lamp was installed opposite the Corner Works (as it became known) but there is no evidence that Malcolm had gas laid to it because he was already generating his own electricity, albeit at 110 volts DC, for both his house and the Corner Works. For the former he made use of a set of accumulators and 110v. DC motors drove line shafting for the machines in the Works, which had increased in numbers.

The mid 1920's saw the beginning of the disastrous slump in agriculture when the prices of farms, livestock and deadstock more than halved and many farmers went out of business. There was very little money about and lots of belt-tightening had to take place. Consequently the steam engine builders suffered from the lack of orders and also, due to the advent of the modern internal combustion engine, saw a lot of their business taken over by Tractor manufacturers. Despite amalgamations they never recovered although some respite was given by increased demand as a result of World War II. During this period the farm horse was still much in use and this provided Bloomfields with a considerable amount of work, repairing wagons, tumbrils and other associated equipment.

*Garrett 4CD tractor, fitted with a canopy and Boulton blocks on the wheels for road work. Taken c1928 at Bloomfield's Works. l to r Charlie Clarke, Roy Bloomfield, Donald Martin, Malcolm Bloomfield.* David Kindred Collection.

Around this time Malcolm also had the foresight to enter the world of factoring and he acquired agencies from firms such as Garretts, Ransomes,

*Ruston Proctor traction engine 'Jack' at Mr Feveayear's farm, Wingfield, c.1960 when owned by Mr Whipps of Bungay. He is on the engine and Malcolm Bloomfield is in the centre wearing a cap and tie.* Neil Bloomfield Collection.

Massey Harris, Barnlett, Bomford, Bentall, Albion, Hunts and others, which allowed him to sell a wide range of machines and implements.

He also became an agent, to sell the small engines of Lister and Petter and, the larger lamp-start paraffin and cold-start diesel engines of Blackstone and Crossley.

Bloomfield's business had now grown such that it had several employees and the whole corner site, with the exception of the house, was occupied by the firm. Also Malcolm was travelling around the countryside, firstly on a "Levis" motor cycle then in a "Standard" car, buying used steam engines, boilers, threshers and industrial equipment and taking them back to the Corner Works for refurbishment and repair before selling them. Consequently more space was needed and c1930 part of the "Broad Meadow", just around the corner

*Burrell compound traction engine No.3925 owned by R.H & R. Paul, in the yard of Bloomfields* David Kindred Collection.

from the works, was bought from Herbert Gooding of Hill House Farm who ran a nearby saw-mill, operated by a gas engine. This piece of land was then fenced off and buildings were erected on it. The site was named "Meadow Works", as it is today, and it afforded a little space for a garden to enable Malcolm to take his mind off the business and grow vegetables. A sty was also erected and pigs kept there, providing meat for the family and neighbours and also fertiliser for the garden.

A small Crossley petrol/paraffin generator, with a 110v. dynamo to provide the required lighting, drove a bandsaw for cutting apple-wood cogs for mill gears and poplar wood bands for steam boiler lagging.

At the Meadow Works threshing machines, seed hullers, chaff cutters and straw pitchers, along with other wooden equipment such as wagons, carts and trailers were repaired; steam engine repair and refurbishment was still carried out at the Corner Works.

In the early thirties, because of the recession, Malcolm sought other business and he made a very important move. At this time, electricity and diesel power units were replacing steam engines and the early

*Engines at Bloomfield's Corner Works* Neil Bloomfield Collection.

*A good customer of Bloomfields - Day of Coddenham, in front of St. Peter's Works of E.R. and F. Turner of Ipswich.*

Neil Bloomfield Collection.

slow-speed oil engines and Malcolm saw an opportunity to export these machines and some stationary boilers to the Middle East. He therefore scoured nearby counties for these redundant engines and began a brisk business in taking out, returning to Debenham, refurbishing, packing and despatching these machines abroad. His main customer was Geo. A. Bust, an engineering and machinery dealer of Baghdad who purchased a very large number of engines. These were both lamp-start and diesel, which were popular for their slow speed and low maintenance requirements.

*Burrell "Princess Mary" whilst owned by Bert Stock. Taken on July 8th, 1934, at Ipswich.* Kevin Swann Collection.

An example of the above type of transactions is as follows: One evening, around 1934, a knock on Malcolm's door revealed a man who introduced himself as Mr. Reid, the general manager of a sawmill in Borneo. He said that he wished to buy for his mill a redundant Garrett CCS9 steam engine which was currently installed in Snape Maltings but was surplus to requirements. He asked if Bloomfields would dismantle and remove it, convert it for wood burning, then pack and dispatch it to Borneo via East India Docks. On arrival in Borneo it would be unloaded onto a raft and towed several hundred miles up river where elephants would then haul it through the jungle to the sawmill.

During his short stay with the Bloomfields, Mr. Reid recounted several amusing tales. One was that his sawmill was so remote from civilisation that his native workers would often ask for playthings such as footballs, in lieu of wages.

Work on the large Garrett engine was quite a daunting task for such a small firm as Bloomfields especially so as the steel chimney itself was 80 feet high. This was lowered with some tackle which was kindly loaned by Mr. King, the Works Manager at Garretts of Leiston, who was a friend and colleague of Malcolm. With his help this job was successfully completed and Mr. Reid, fully satisfied, paid Bloomfield's bill promptly.

One of Malcolm's interesting jobs in 1930 was the overhaul of three Showmans engines which arrived together. They also all required the replacement of their front and rear wheels with rubber tyres. This job provided quite a lot of work for the firm at a time when it was needed.

During the latter part of the 1930's and early 1940's the general economy improved such that people wished to take advantage of electricity as a form of power and the firm took to supplying electric generating sets, mostly direct current but in later years the standard alternating current versions. Farms, shops and large country houses were some of the places that Bloomfields equipped with electrically driven water pumps, lighting equipment and various small machines. Back up for lighting was provided by banks of open-topped glass accumulators, the average size being 18 inches high by 9 inches square. These were charged when the engines were running but, because of their design, every two to three years they had to be taken down from their wooden trestle containers, emptied, cleaned and refilled with a mixture of water and sulphuric acid. This was a filthy job and many sets of overalls were ruined by acid splashes. This was also hard work as about sixty of these accumulators on a 110v. supply had to be processed in one day.

*Garrett superheated tractor No.32647 owned by R.R. Harvey of Tannington, under repair at Bloomfield's Corner Works.* Photo: R.G. Pratt

Most people by now owned a wireless set run from a battery and accumulator and Bloomfields, as well as garages, bicycle shops etc. provided a re-charging service for about sixpence a charge. Each wireless-owner had two accumulators one in use and one being re-charged.

The Bloomfield family had by now grown considerably and, keeping up with the times, Malcolm bought a large 110 volt electric oven made by Westinghouse of America. This had to be run by the engine and so of course, due to the family's principles, it was never used on a Sunday. Malcolm also purchased from America, in 1935, as a novelty, a 'Schick' 110v. DC electric shaver. This is still in the family's possession in its original box.

In 1938 the small Blackstone engine driving the Corner Works equipment was replaced by a new Crossley with twin 5 feet diameter flywheels. This was a very nice engine which was started by compressed air at 7 a.m. each morning and, because of a shortage of space, was installed underground in a glazed- brick engine house, sealed with asphalt. Gibbons of Crowfield were the builders, the asphalt company was Passini of Ipswich and the ceiling of the engine house was supported by 8 inch by 4 inch RSJs. This engine had a welding DC generator on the left hand end of the crank and on the right hand end there was a similar DC generator to supply the other machines and for charging. Needless to say it was often overloaded and was overhauled every Good Friday. It ran until 1959 when it was sent to Oulton Broad and exchanged for a "Fielding" which was more suitable to export. The engine room, which could have served as an air raid shelter, was finally filled-in when the site was redeveloped in 1998/99.

*A timbered Merchant's house in which, until recently, Rae Potter (née Bloomfield) lived. The large building is the Forresters' Hall.* Photo: Pat Freeman

Electricity from Bloomfield's Corner Works was supplied by an overhead cable across the High Street to the Forresters Hall from the late 1920's until the early 1940's, when mains electricity was connected to the Hall, as it was then being occupied by the military authorities.

*Bloomfield's skeleton wheels, patented c.1930 and improved afterwards.*
Neil Bloomfield Collection.

*Cement mixer made by Bloomfields.*
Neil Bloomfield Collection.

The 1930's saw an increased demand for equipment and tackle and Bloomfields often came to the fore with items of their own design. One thing which gave the firm a steady trade from then, even up to 1965, was the manufacture of steel skeleton tractor wheels which Malcolm devised. They had gripping spuds acting in a similar fashion to a horse's hoof, the idea for which came from watching ploughing horses in action. Bloomfield's skeleton wheels were popular in several counties, especially in the exceptionally dry year of 1959 when the hard ground was ruining the rubber tractor tyres when ploughing. Demand for these wheels was so great that the firm could not keep pace with the orders and it contracted with Garretts of Leiston, and they were pleased to assist, to produce 75 pairs of these wheels to fit various types of tractor. Some skeleton wheels were exported, the farthest to Cambodia. Also 'road bands' were made for some of the early design of wheel and, until 1950, Cambridge rib rolls and flat rolls were manufactured.

*Demonstration of Bloomfield's mole pipe-layer.* Neil Bloomfield Collection.

There was a call for general purpose cement mixers and Bloomfields made many of these, driven by Lister or Petter $1\frac{1}{2}$ h.p. engines. Garretts cast the mixer bowls and the main pinion, and the wood patterns for these are now back at the Leiston Long Shop Museum after they had been carefully preserved by Bloomfields at Debenham.

Another example of the firms creativity and scope of work was as a result of a bad accident at a sawmill at Beccles. The firm had several saw

*Burrell works No.1453 built 1889. Owned by Robert Hines who is on the engine with the oil can. Just below him on the left of the fly wheel is Mr Clark and to the right of him is William Theobold. Taken c.1920 at Town Farm, Thorndon, about 5 miles north of Debenham. This engine was almost entirely re-built by Bloomfields.*

Neil Bloomfield Collection.

benches all operated by belts from an overhead shaft which was driven by a stationary steam engine. If, for any reason, a saw bench had to be shut down (apparently it did not have a system of fast and loose pulleys for engagement) the engine stop-lever had to be operated which was very stiff. As a result of this accident the Factory Inspector insisted that each saw bench should have a push button device to allow the engine to be switched off. Bloomfields therefore designed, wound and fitted a large solenoid, which could be operated by push buttons from any bench, to the engine stop lever to enable the engine to be shut down very quickly.

One of the highlights of Malcolm's career must have been the successful conversion of heavy steam mole drainers "ex-Fowlers" into direct pipe-laying machines. These consisted of a 6 inch diameter mole with a very large shield either side, to keep the forced earth open, and a long, gently curved, tube reaching down into the mole cavity into which standard 4 inch diameter drain pipes would be fed by hand from a trailer running alongside. Unfortunately this was a rather short lived success as the power to haul the equipment was provided by Fowler cable ploughing engines which, although capable of immensely valuable heavy cultivating work, were in their last days as a preferred power unit. However, similar conversions were carried out on lighter mole drainers for the purpose of laying communication cables across airfields during World War II.

*Titan tractor, with Jimmy Mutimer at the wheel. Bloomfields repaired the engine.*

Neil Bloomfield Collection.

Even 'church work' was tackled by the firm and at one stage the bell in Pettaugh church was removed with all its gear, and ball bearings were fitted to the axle and the whole was reassembled.

Another interesting installation took place in 1938 which involved fitting a new 60 h.p. Blackstone engine, together with a Christy & Norris Pulveriser, at Blakenham chalk pits. The firm even built the workman's room, a timber structure which later was removed in one piece and taken to the Meadow Works where it was used as a store. Sometime after the war Bloomfields purchased the engine, reconditioned it and sold it to the agent in Baghdad.

Stationary steam boilers were fitted into local dairies, farms, abattoirs and factories and in the 1930's for some of the larger dairy farmers, the firm installed milking machines. All were engine driven with cascade type milk coolers and improved farm water supplies. There were no bulk tanks as milk churns were still in use. The abattoir work consisted of fitting steam boilers, scrub tubs, meat hooks, hoists, generators and the like and in some instances Bloomfields actually made and installed the cold rooms and the refrigeration plants.

In the late 1930's tractors were becoming more plentiful, mainly Fordson, Case, International and Allis Chalmers, with a few Massey Harris and other imported makes. These all needed frequent servicing including major repairs such as replacing big-end bearings and re-boring cylinders. Bloomfields purchased a portable 110v DC cylinder boring machine, from 'Buma' of Newcastle-on-Tyne, which could be clamped on the cylinder block and each cylinder was bored out individually. This machine is still in the family's possession.

*Garrett tractor No.34641, once owned by M.Bloomfield who sold it to H.H. Heffer of Farnham, Suffolk. Taken on August 25th 1934. Now called 'Bunty' and owned by the Bressingham Steam Museum.* Neil Bloomfield Collection.

Local millers provided quite a lot of work for the firm as from time to time, there were mill engines to install and these together with other milling machinery, sack hoists etc. required repair and maintenance. Power to drive these items was from overhead shafting via flat belts of which the firm still holds large stocks.

The bulk of the firm's work, up until the Second World War, was concerned with the supply, refurbishment, repair and maintenance of steam engines and associated equipment, including conversions.

Surrounding Debenham were many firms, some very small, which carried out contract threshing and ploughing for local farmers and they provided Bloomfields with a flow of work which was mainly done at the Corner Works.

*A Burrell workbench in Bloomfield's workshop. This was acquired about 60 years ago from Garretts who got it from the Burrell works.*
Photo: Pat Freeman

As one might expect, due to the proximity of Debenham to Leiston, there was a preponderance of Garrett engines for Bloomfields to look after and they had a very good working relationship with the Leiston firm up until the time that Garretts sadly declined in the early 1940's. 'Bunty' Works No. 34641 was a Garrett 4CD tractor that Bloomfields worked on in the late 1920's. This engine had been sold to the Clare District Council in 1925 but just after having been fitted with rubber tyres in 1928, a disastrous fire broke out at the Council's depot and Bunty was badly damaged. The charred remains were sold to Bloomfields who restored it mechanically, made and fitted a new canopy and gave Bunty a new lease of life with H.H. Heffer, Agricultural Threshing contractors of Farnham, Suffolk. This engine can now be seen at the Bressingham Steam Museum, Norfolk.

Another source of work was from Showmen and amongst Bloomfield's customers were Bert Stocks of Ipswich, Barkers of Ipswich, F. Harris of Clare, W. Wright of Newmarket and Thurstons of Norwich.

Conversions, although now frowned upon by many members of the preservation movement, were not uncommon and Bloomfields became involved in this aspect of the business. An example of a conversion involved a Garrett 4CD compound Tractor No. 33358 which was War Department surplus and was acquired by Bloomfields in 1921. This machine was converted to a Showmans engine and was then sold to F. Harris. The engine, called Lord George, is now owned by Kevin Swann,

*Garrett Showmans engine 'East Lynne'. This was used in the Leiston area with a travelling circus. It was made at the Garrett Works c.1911 and Malcolm Bloomfield, then an apprentice, helped to build it.* Neil Bloomfield Collection.

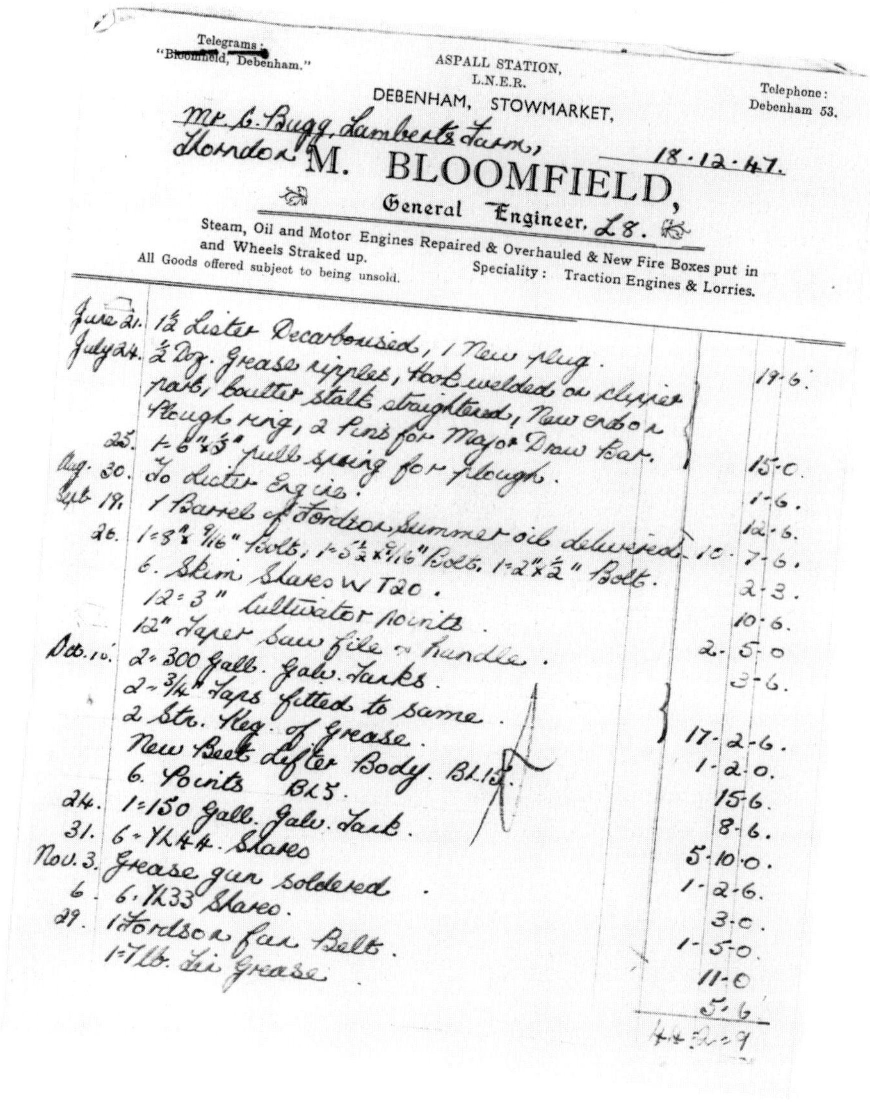

Telegrams: "Bloomfield, Debenham."

ASPALL STATION,
L.N.E.R.
DEBENHAM, STOWMARKET,

Telephone: Debenham 53.

Mr C. Bugg, Lamberts Farm,
Thorndon 18.12.47.

# M. BLOOMFIELD,

General Engineer. L8.

Steam, Oil and Motor Engines Repaired & Overhauled & New Fire Boxes put in and Wheels Straked up.

All Goods offered subject to being unsold.

Speciality: Traction Engines & Lorries.

June 21. 1½ Lister Decarbonised, 1 New plug.
July 24. ½ Doz. Grease nipples, Hook welded on clipper
part, Coulter stalk straightened, New end on
Plough ring, 2 Pins for Major Draw Bar.
25. 1-6"x½" pull spring for plough.
Aug. 30. To Lister Engine.
Sept 19. 1 Barrel of Fordson Summer oil delivered.
26. 1-8"x 9/16" Bolt, 1-5½x9/16" Bolt. 1-2"x½" Bolt.
6. Skim Shares W T20.
12-3" Cultivator Points.
12" Taper Saw file & handle.
Oct. 10. 2-300 Gall. Galv. Tanks.
2-¾" Taps fitted to same.
2 Str. Keg. of Grease.
New Beet Lifter Body. BL15.
6. Points BL5.
24. 1-150 Gall. Galv. Tank.
31. 6-YL44. Shares
Nov. 3. Grease gun soldered.
6. 6. YL33 Shares.
29. 1 Fordson Fan Belt.
1-7lb. Li Grease.

19-6.
15-0.
1-6.
12-6.
10-7-6.
2-3.
10-6.
2-5-0
3-6.
17-2-6.
1-2-0.
15-6.
8-6.
5-10-0.
1-2-6.
3-0.
1-5-0.
11-0
5-6
44-2-9

Chairman of the East Anglian Traction Engine Society, having been passed to him by his father, George, who purchased it in 1954.

It was customary for the couple of dozen or so Threshing Contractors close to Debenham to send their engines and Threshing Drums to Bloomfields, at intervals of about two years, for overhaul. The engine drivers and mates on these occasions would travel each day from their home by cycle to Bloomfields to assist with the overhaul which normally lasted about two weeks and was probably as near to a holiday that these men ever had. Consequently their lunch hours were mostly spent at the nearby Cherry Tree Inn and often their reliability in the afternoons was not too special. It must be said, however, that, whilst perhaps lacking a good education, most were very intelligent, hard working, conscientious and of good character. Their working day was longer than most, having to start very early to raise steam and they generally had to clean out the engine boiler on a Sunday. Also they were responsible for seeing that the machinery was in good order on a day to day basis and had to keep account of the work they carried out.

Whilst the Corner Works were kept busy attending to the engineering side of the business there was no lack of activity at the Meadow Site. A constant stream of work of all sorts kept the staff there fully occupied. The blacksmith, wheelwright and general engineer would co-operate on 'shoeing' days. These happened about four times each year, after the wheelwright had rebuilt several wooden wheels with new spokes and felloes. Huge fires were built to heat the metal tyres, either the originals or new ones, which were then shrunk onto the wooden wheels which, if not promptly cooled, would start to burn. After several years using the time honoured way of carrying out this operation, Malcolm contrived a sunken pit filled with water, with the shoeing platform suspended by spiral runners so that as soon as a red hot tyre was properly located on the wooden rim the whole lot could quickly be lowered into the water for cooling. The wheel would then be raised very quickly for final inspection before being lowered again to be fully cooled. This device saved a considerable amount of time but was only used for a few years, until Bloomfields ceased shoeing.

# World War II and Post War Years

The onset of World War II had a great effect on businesses and brought about much social change as well. Young men and women either volunteered or were conscripted to serve in the Forces and this caused a severe labour shortage.

Increased food production was demanded by the Government and this effort placed a great strain on all aspects of machinery used in farming and market gardening, thereby leading to much work for Bloomfields. Luckily for the firm, at the outbreak of war, five of Malcolm's seven sons were working with him, four of these had "exempt occupation" status and the fourth was only fourteen. The other two sons were nine and three. These six family members together with the firm's general engineer, the wheelwright and the blacksmith who were all over military age had to cope with all the work at Bloomfields. They did have the occasional help of one or two pensioners and a younger employee who later had to work as a "Bevin Boy" down the coal mines.

*Garrett tractor BJ 1722 owned by Henry Capon, of Kenton Hall. Taken at Cross Green, opposite Bloomfield's Corner Works, in 1918.* Neil Bloomfield Collection.

Great strain was placed on this work force, not only by the increased work involved but because of the civil duties they performed, such as being in the Home Guard or in the Debenham Fire Brigade. A typical day might be from 7 a.m. to 9 p.m. in the Works then, on five nights per week, sleeping in a hut close to the Meadow Works in readiness for night call-out for fire duty which was quite frequent, as were day calls when all work was left so as to attend fires over a wide area.

Incidentally, the firm provided an air raid siren for Debenham, using an old steam whistle attached to the large compressor tank in the Corner Works. Instructions for 'Raid' or 'All-Clear' were received over the telephone and actioned by whichever member of the family took the call.

Perhaps there are not many people who can recall the sight of huge numbers of traction engines, portables, threshers and other pieces of heavy machinery placed at intervals over some of the larger fields in the countryside. This action was taken by the Government to prevent aircraft, including gliders, from landing successfully. Where on earth so many machines were found in such a short time is something on which to speculate; doubtless today they would be worth a small fortune to preservationists.

*Garrett tractor No.34359 at a sale on July 12th 1950. Owned by H.Cotton & Son of Swilland, it was sold to Williams of Mendlesham.* Neil Bloomfield Collection.

Scrap iron was urgently needed to fuel the war machine and gates, railings and all sorts of iron work were confiscated for this purpose. Bloomfield's scrap was generally of good quality and was carted away by Hales, of Beccles, using a Lanz Bulldog tractor or a Clayton and Shuttleworth steam wagon.

As petrol was rationed, only business journeys for cars were allowed and from time to time one or two farmers, anxious for a bit of social life in Debenham, would make the excuse of going to Bloomfields to collect plough shares. Needless to say, in a few cases, the same half dozen plough shares occupied the boot of some cars for the duration of the war.

The Leiston firm of Garretts was sadly in decline as far as engine building was concerned and in the early stages of the war they offered Bloomfields their remaining stock of new, but unfinished, steam boilers. Some of these were then disposed of to small dairies, laundries and market gardens. Also, when the American controlled airfields came into being around Debenham, engineers would come to Bloomfields for all types of steam boilers and tanks, to cater for their hot water needs.

Garretts, because they were heavily involved with munitions work at their Leiston factory, were unable to look after the maintenance needs of local engine owners and Bloomfields acquired more work because of it. One such job was the overhaul, straking and making from scratch a new firebox, for Garrett Tractor 32944, owned by Abbotts of Saxtead. This engine was later owned by Billie Jeanes and is now in the possession of the Leiston Long Shop Museum.

*A one ton electrically driven Portable Meal Discharge Hopper c.1965 designed and made by Bloomfields.*
Keith Bloomfield Collection.

Every few years boiler tubes had to be removed from the engines to remove scale so as to maintain efficiency to raise steam. This had always been done by hand, was very hard work and labour intensive. Therefore, because of the labour shortage the eldest Bloomfield son, Roy, designed and built a machine to do this work. It was very successful, although exceptionally noisy as the jingling of the hollow tubes passing through the cleaner could be heard up to a mile away.

As the war drew to its close a few restrictions were lifted and life was a little easier. However there was, by that time, a tremendous amount of work coming to Bloomfields and the Post War era held great promise for the firm.

As mentioned elsewhere, Combine Harvesters were first sold by the firm in 1946 and this marked the start of more work for Bloomfields.

There followed a constant demand for systems to handle bulk grain from trailers, such as driers, cleaners, silo storage and conveyors. Starting with plants that coped with about five to six tons per hour, a gradual increase has occurred up to the present day 60 - 100 tons per hour.

Bloomfields have made thousands of grain pits of all sizes and have manufactured and assembled elevators, trough augers, hoppers and a host of specialist equipment, all for the purpose of handling grain. To this day elevators and silos are still being supplied and installed.

Only a few milling plants were in place before the war but since 1950 the firm has installed several hundred electrically - driven plate mills, hammer mills, roller mills and structured mills with fluted rollers, with vertical and horizontal mixers from a quarter to two tons capacity. Thousands of spiral grain augers, rigid and flexible, have been supplied for grain handling and livestock feeding. Many of the plants installed by Bloomfields have been designed and built by them to suit the requirements of individual clients and are therefore unique to the Debenham firm.

One aspect of Bloomfield's work that has been constant from the 1950's to the present day is the supply and installation of livestock feeding equipment. These systems, in many varied forms, consist of feed hoppers up to 30 tons capacity, either purchased or made by Bloomfields and conveyors by auger, chain, rope or disc operation. These systems are supplied to piggeries, cow sheds, deep litter houses for hens and ducks, hen batteries and bullock yards.

The firm devised a considerable amount of work from contractors who laid mains water to the village after the war and later when main sewerage was provided.

Between the mid 1950's and the 1980's Bloomfields sold thousands of implements to local farmers such as ploughs, harrows, drills, fertiliser and muck spreaders, balers, beet harvesters, elevators, rotovators and the like. Also from 1954 until 1967 they sold Ferguson tractors in large numbers and some Fordson and other makes. Although everything on the farm and in small businesses is nowadays powered by mains electricity, it is worth noting that almost five hundred tractor-driven stand-by generators have been sold by Bloomfields over Suffolk, Norfolk and Essex.

*A wire-layer using a mole; made by Bloomfields for Eastern Electricity.*
Neil Bloomfield Collection.

The headed notepaper used by the firm up to about 1950 mentions 'Aspall Station, L.N.E.R. Debenham, Stowmarket' which was the rail facility on the mid-Suffolk line that Bloomfields used up to 1952 when the station closed. For a few years they then used first Haughley station then Stowmarket depot until British Rail ceased local deliveries. Since then the firm has relied almost totally on road transport.

It is interesting to note that Bloomfields purchased Kenton station in 1954 which was two and a half miles from Debenham but, having no use for the site, they sold it for £150 to E.P. Clarkes of Bedingfield.

The year 1962 saw the almost complete closing down of the Corner Works as the Meadow site was enlarged by the purchase of adjoining land and more works buildings and offices were erected on it. Corner Works was retained as a useful store until 1997 when it was sold for redevelopment.

Today the firm has three directors, two managers and an assistant, nineteen fitters and two regular office staff.

A comprehensive catalogue is available that covers the operations of Bloomfields which can be summarised as follows:

- Sale, repair, maintenance and hire of Combine Harvesters and ancillary equipment. Making trailers to transport the headers of Combine Harvesters.
- Supply, repair, maintenance and hire of a wide range of equipment for the smaller builder, horticulturist and gardener.
- Repair and supply of parts for tractors.
- Fabrication of special gates, railings and other iron and steel work of all shapes and sizes including farm and house names in ornamental ironwork.
- Holding large stocks of spares of electrical, engineering and agricultural equipment, general tools, ironmongery, hardware and water supply related products.
- A wide variety of work is carried out on farms and light industrial sites.
- The supply and maintenance of Livestock feeding equipment.

In effect, the family firm of Bloomfields, in existence for the best part of one hundred years, caters for the majority of needs of the farmers and agriculturalists who live around Debenham and it is therefore an essential part of that rural community.

# M. Bloomfield & Sons – Personnel

M. Bloomfield & Sons Ltd. is a fully independent family-owned firm whose standards were set by the founder, Malcolm Bloomfield and have been upheld over the 88 years of its existence by a dedicated staff whose aim has always been to deal to the satisfaction of the client.

Malcolm and his wife had a large family of seven sons and seven daughters and all the sons have been, or still are, part of the firm and several of the daughters have had some association with it.

Mrs. Bloomfield was a typical country wife whose family meant everything to her and her support of Malcolm in all his business matters was unquestionably loyal. Her purpose in life was to bring up the family and keep house (and for a period the firm's books) and she did this purposefully and cheerfully until she died in 1971.

Malcolm, as head of the family and the firm, was a sound adviser who always encouraged his family to have a shot of any business matter. He had a pleasant, natural way of doing business and this manner established for him and the firm an enviable status of trust which was widely acknowledged in East Anglia and even further afield.

In 1951 a formal partnership was made between Malcolm and

*Wallis & Steevens Reg No.7871, bought by Neil Bloomfield on July 7th 1956. He sold it to a Mr Mason c.1966. Taken at the Woodton rally in 1958. Standing on the left is Keith Bloomfield, then Neil. The lad on the engine is Alan Hines with his father.* Neil Bloomfield Collection.

his seven sons which operated smoothly until 1962 when sadly Felton (aged 42) and Hilton (aged 27) were forced to retire owing to ill-health.

M. Bloomfield & Sons then became a limited company with Malcolm, Roy (aged 50) Cecil (aged 49), Russell (aged 44), Neil (aged 37) and Keith (aged 32) as Directors.

During the time working with the firm each son had his own area of main responsibility:

Roy dealt chiefly with traction engines and heavy oil engine repairs.

Cecil concentrated on small engines, electric lighting plants and electric installations for farms, shops and large houses. He was an expert on hay machinery and self binders for corn.

Russell was initially in charge of the woodwork side of the business, looking after the wheelwright and carpenter. He was then involved with electrical plant and built up a useful business in electric motor servicing and replacement. The Company now holds a stock of about 250 AC type motors from 1/4 h.p. to 40 h.p. Russell also became heavily involved with grain handling and milling work up to his retirement in 1990.

Felton did much lathe work and operated a large planing machine.

Neil has run the Combine Harvester, Baler and Hay machinery department almost exclusively since 1951.

Keith joined the firm in 1944 mainly to concentrate on grain handling equipment and power transmission. In 1951 he did his National Service and in 1954 he combined with Russell looking after grain handling and milling work, as well as acting as Company Secretary and running the financial and administration side of the business.

Roy died after an operation in 1983 and Felton, after a life time of ill health, passed away in 1994. Cecil who carried on with the firm into his early seventies retired and passed away in 1996. Hilton, the youngest of the family, struggled with bad health all his life and finally succumbed in 1997.

Russell retired in 1990 and the present Directors are Neil, Keith and Keith's son, Simon, who took over, at Russell's retirement, the responsibility for the grain handling and milling work which includes the busy steel fabrication and special unit work.

Some of Mr. and Mrs. Bloomfield's daughters also played a part in the firm's activities. Eva worked there for a short time before joining the WAAF during the war, Rae spent some years running the office, Brenda was part of the firm until she succumbed to a brain tumour at the age of twenty five and Rita helped out in the office but was happier doing outside work for the firm. The eldest sister, Marion, must not be forgotten for although she did not actually work for the firm she played a very important part in enabling her father and mother to spend extra time on business matters. Marion worked from the age of fourteen until she was about sixty, looking after the household and, in their early life, caring for the younger Bloomfield children as well as looking after her own family.

To be successful, a business cannot run smoothly without dedicated staff and Bloomfields are no exception and over the decades of its existence it has been blessed with a number of notable employees, four of whom have since passed away.

Ernie Rozier, who as a boy had committed some youthful misdemeanour and had been punished for it with a short spell of detention, stood outside the Corner Works one day looking at what was going on. Needing a lift with a heavy engine part Malcolm shouted to him, over the fence, for his assistance. Ernie willingly went in and gave a hand. Fifty years later when he retired from the firm and tributes were made to him, his faithful service, adaptability and contribution to Bloomfield's success over those years was suitably acknowledged by everyone who had worked with him.

*Cross Green, Debenham, with St. Mary's Church in the background. Just in front of the Church is Bloomfield's Corner Works.* Keith Bloomfield Collection.

Alfred Muttick, the blacksmith, came to help out for one week and stayed for thirty one years. A veteran of the Boer War, he was a man of great probity and was for many years a member of the church choir. Young employees of the firm and even the sons of the Bloomfield family dare not address him other than Mr. Muttick. Although at times his manner was a little peremptory he was always unfailingly courteous to Mrs. Bloomfield.

Reg (Dick) Brown was one of the firms first electric welders. Malcolm had made his own portable electric welder around 1930 that was driven by an old Morris petrol engine (it continued in use for some

*Ready for sale. Herbert Hawes, of Stradbroke, a customer of Bloomfields, has engines lined up for sale in the 1950s.*

Neil Bloomfield Collection.

twenty years) and on this Dick received some instruction from Malcolm and learned the rest as he went along. He could do almost anything and was particularly noted for overhead welding in traction engine fire boxes where he spent many hours. He was a man of dry wit and humour and remained a faithful servant for some 48 years. He passed away in January 2000 at the age of 88 years.

William Denny was a skilled wheelwright who could tackle any wheel including the tricky small diameter front wheels of a thresher, as well as the light ones used on butcher's carts, and wagon and tumbril wheels. He rebuilt a few burnt-out elevators using oak for the frames and Columbian pine for the trough boards and also re-used the iron work. He was a master at making thatching ladders, often going to the wood importers to select the baulks himself. He cycled to work six days a week for about thirty five years. He was a dour man, determined to do things properly without waste of materials, with as little time lost as possible. He referred to idlers as those who "dragged a light harrow". Unfortunately he passed away soon after retirement in the early 1950's.

There are several employees still with Bloomfields who have completed over thirty years of service. They include Derek (Mick) Barker, who joined as an apprentice in 1949 and although officially retired still works on a part-time basis, David Whiting with over 40 years service, Jeff Mutimer with 42 years service and Stewart Mullinger who has been with the firm for 31 years. These four joined the firm straight from school. George Stevenson came to Bloomfields in 1954, had a short break elsewhere, and then returned and is still employed there.

Several others of the present employees joined the firm straight from school, including David Webb, Robert Scase, Mark Chilvers, Duggie Davey, Ian Bloomfield (whose father worked for the firm in the 1940's), Paul Cattermole and Glen Harrison. The firm has always had a policy of employing young workers because, in the words of Keith Bloomfield, "If you don't have boys you won't have men" and generally several are with the firm at any one time through school release and work experience schemes.

It should be mentioned that Helen Keeble has been at Bloomfields working in the office for the last 26 years.

Bloomfields is proud that it is very much a family firm, the term "family" extending to, and including, the employees. This unique status typifies the character of country folk in creating and maintaining a satisfactory and enviable working environment.

# Combine Harvesters

Combine Harvesters were first sold by Bloomfields in 1946 and these were allocated by the War Agricultural Committee, but on what basis the firm did not know.

The first one was a Massey Harris No.15, an 8' 0" trailer model pulled by a tractor, but with its own American engine, possibly a "Continental'. It was obtained directly from Canada, arriving at Stowmarket Railway Station in its own crate which was taken to Bedingfield Hall Farm and assembled there. Incidentally many of these crates survived over the years as sturdy garden sheds.

The second and third that arrived that year, also directly from Canada, were self-propelled 12 feet cut Model 21, Massey Harris Bagger Combines with 6 cylinder Chrysler petrol engines and they cost £900 each. One, assembled in Bloomfields yard, went to Crow's Hall, Debenham and the other, assembled on site, went to Reginald Harvey at Braiseworth Hall, Tannington.

It is no secret that steam threshing men treated the introduction of combines with a derisory attitude, saying they would not last long, but within two years they had all but taken over from steam threshing.

This was because of the economy of harvesting that the combine could offer. It was capable of harvesting up to 20 acres per day (say 40 - 50 tons of grain) using only four men, two on the Bagger Combine and two carting the grain right into the store. The fuel costs were low as they used only about two gallons of petrol per acre.

It was not long before Tanker Combines came on the scene,dispensing with the need to pick up the heavy bags of grain, weighing about 18 stone, from the ground. This method of harvesting required the installation of grain reception, cleaning, drying and bulk storage equipment with which today Bloomfields is still heavily involved. However, the firm still sold a few Bagger Combines as late as 1956.

It is surprising that even the early models of Combines could satisfactorily handle a wide range of crops such as wheat, barley, oats, peas, beans, trefoil and clover seeds. Nevertheless, always seeking perfection, Bloomfields replaced the separating rotary screen on the Bagger model with a Penney & Porter adjustable screen, as used on a convertible threshing drum.

1948 saw the introduction of the first small Massey Harris self-propelled 8 ft. cut models 222 and 722 and in 1949 Bloomfields were dealing with the British made model 726 S.P. Combine from the Massey Harris Kilmarnock factory at which Keith Bloomfield had spent some time during 1948. These were fitted with either Austin 6 cylinder engines or Nuffield 4 cylinder engines, still running on petrol. However, Vaporiser conversion manifolds were made locally so that, when fitted, these engines could be run on paraffin, thereby effecting another economy of operation.

In 1948 Malcolm Bloomfield designed and built a Straw Press and Knotting Mechanism which went on the back of a model 21 Combine and he developed this with another prototype which was successful. This was never pursued as the Rausendorf Press from Scandinavia was widely marketed and sold in large numbers.

*This Massey Harris Combine Harvester was sold by Bloomfields to F.E. Day and Son, Coddenham, in 1954. In no particular order: Terry, Mrs Bloomfields pet, Neil & Keith Bloomfield, Bernard Race, Derek Barker, Peter Hailes and George Stephenson.* Keith Bloomfield Collection.

After 1950 there were no further imports of the S.P. model 21 Combine as Massey Harris then produced, at Kilmarnock, model 780 Combines with a 10 feet or 12 feet cut. Most of these machines were fitted with diesel engines, possibly Perkins L4.

Massey did produce a 735 Combine around mid-1950 to 1960 that was self-propelled but it did not sell in large numbers.

All the early Combines had a weakness in that the cutting knife had difficulty in coping with British crops which were generally heavier and damper than those in America. Yet again, Bloomfields made a successful modification; in the form of a smaller Vee pulley to give the knives more speed.

Around 1960 "Guarding" Rules and Regulations were introduced and, as a consequence, Bloomfields made and sold over a wide area several hundred sets of safety guards for Massey Harris Combines. Throughout the above period Bloomfields were supplying Massey Harris pick-up balers, both Power take-off driven or with various built-in engines such as the American'Wisconsin', the Coventry Climax and Armstrong Siddeley diesels.

*A New Holland Combine Harvester Model TF78. It has a 30ft cut.* Photo: Keith Bloomfield

The last Massey Harris Combine models that Bloomfields sold were in the '500' range. In 1962 the firm switched to selling Claas Combines and New Holland Balers. In 1966 New Holland (now Ford New Holland) took over the Clayson Combine factory in Belgium and from then on Bloomfields sold only Clayson Combines and New Holland balers.

The conditions of operation of combines have changed enormously over the years and at the present time, within air-conditioned drivers cabs, there are complicated electronic controls for the hydrostatic devices, shaft speeds, travelling speed, cutter table height, straw chopper, chaff spreader and grain loss monitor. The driver therefore had always to be on the alert whilst operating these huge machines but he does have a radio/cassette player to listen to.

Combine models from 1966 to the present day have been many, with horse powers ranging from 60 H.P. to 360 H.P. and the top of the range model is capable of harvesting about 100 acres per day. This machine costs in the region of £200,000, which seems a huge amount of money considering it is used for only about three weeks of each year between late July and August. However, such are the vagaries of the British weather that these high capacity machines are accepted as necessary.

Technology is now available which allows precision farming to be taken almost to its limits. Combines can be fitted with Crop monitoring and Yield mapping systems which, when allied to accurate continuous satellite plotting of the combines position whilst harvesting, will provide invaluable data to enable the farmer to efficiently husband his land.

Bloomfields, as well as selling new and used combines, also hire them out for the harvest season.

# Debenham - The Sign

At the southern approach to the High Street there is a splendid coloured sign, the constituents of which succinctly sum up the history of this ancient and picturesque village that lies in the heart of the lovely Suffolk countryside.

The main breadth of the sign, five feet wide, depicts a bridge over running water which is a reminder of the origin of Debenham which stands at the confluence of four little streams and marks the effective start of the river Deben.

*Debenham Village sign, made by Keith Bloomfield.*

Photo: Pat Freeman

It was along this waterway, much wider and deeper then, that some Angles must have sailed from the east, in the 6th. century. They chose to settle at the head of the river and gave the site the name Debenham, derived from the Anglo-Saxon meaning a deep and low-lying enclosure.

It lies in the centre of many square miles of heavy clay land and the settlers appear to have facilitated its drainage by digging moats which provided a network of defences and also allowed them to start "fish farms".

The Gilded Heraldic Ram at the top of the sign denotes that Debenham was one of the wool towns, whilst the wheat sheaves on either side refer to its long association with agriculture. There are many outstanding, ancient buildings in

*Water Lane, Debenham.*
Photo: Pat Freeman

*The High Street, Debenham looking south. Baldry Bros. Grocers, is on the left.*
Neil Bloomfield Collection.

Debenham two of which St. Mary's Church and the Guildhall are depicted by cut-outs which are silhouetted on top of the sign, either side of the ram. The sign commemorates the Silver Jubilee of Queen Elizabeth II and it was made by Keith Bloomfield to a design by the late Tony Williams, a local Architect; it was unveiled on October 14th. 1978.

Keith spent a whole year making the sign which is mainly of mild steel with the word Debenham and the wheat sheaves made in non-corrosive aluminium. A local timber yard supplied a rough-sawn post, obtained from a Suffolk grown oak tree, and it was fashioned and finished by Keith.

*Page's Mill, on the Kenton Road, Debenham.*

Neil Bloomfield Collection.

# This and That

Malcolm Bloomfield shared lodgings at Leiston with a fellow apprentice and it was common in those days for landladies to feed their lodgers as cheaply as possible, especially with rabbit meat. After a prolonged diet of rabbit, Malcolm's room mate became utterly fed up with this fare and absconded one night. In the morning the landlady found a notice pinned on the pantry door which read as follows:

Rabbits young and rabbits old
Rabbits hot and rabbits cold
Rabbits tender and rabbits tough
I've gone home to mother cos' I've had enough.

~ ~ ~ ~ ~ ~

Malcolm remembered a prolonged drought in 1921 when there was hardly any rain from June to Christmas. This, when it did arrive, came in the form of snow and it is said at Nesling Farm, Bedfield, that the snow was so precious that Mr. Robert Nesling ordered his men to sweep up as much snow as was possible and push it into an empty pond.

~ ~ ~ ~ ~ ~

Keith Bloomfield remembers that as a schoolboy in the early war-years he was playing at the end of the railway embankment which was intended to run from Kenton to Westerfield when, seemingly out of nowhere his head schoolmaster appeared and told him to clear off. It was only recently that Keith found out that there was a Churchill's Secret Army bunker underneath that railway embankment.

*An ancient Police House in Water Lane.* Photo: Pat Freeman

~ ~ ~ ~ ~ ~

John Smoothy, great-grandson of the founder of W.G. Smoothy & Son, bought his first car in December 1954 and when he was learning to drive the "qualified driver" with him was his father Norman. Although a "qualified driver" Norman had never driven a car but he did hold an "All Groups" licence which was obtained before Driving Tests were introduced. He was, of course, an expert at driving all types of rollers and steam engines and later he learned to drive a tractor - but never a car ! However, John passed his driving test at the first attempt.

Incidentally John is the proud owner of a classic car, a Vauxhall Wyvern which was first registered on November 2nd. 1950, which he purchased on June 9th. 1989 with 112,760 miles on the clock. John uses the car, Registration number LLR 922, all the year round (except when there is salt on the road) and he can be seen with it regularly at rallies including appearances at Classic Car Shows at the Alexandra Palace and the National Exhibition Centre.

*The old Market Cross Endowed School in the High Street, Debenham* Photo: Pat Freeman

~ ~ ~ ~ ~ ~

The following is a précis of an agreement made on October 21st. 1920 between the Debenham Gas Co. Ltd., 50 Cannon Street London and Debenham Parish Council:

*"The Company contracts to supply the Parish of Debenham with the public lamps of an illuminating power of fourteen candles from October 1st. to March 31st. for the sum of £3. 1s. 9d. per lamp and each will be kept burning from one hour after sunset until eleven o'clock p.m. save and excepting on four nights before the night of each full moon and the night of and two nights after each full moon".*

A Mr. Felton ran the gasworks and he was the lamplighter and lamp extinguisher for Debenham. He was affectionately known as "Gassy Felton".

*Oops! A Smoothy ploughing engine working for Mr Raynor Great Wakering, July 3rd 1930.*

John Smoothy Collection.

*A group of Debenham Gentlemen.*

Neil Bloomfield Collection.

~ ~ ~ ~ ~ ~

Most village people in the early days of the last century would get about on a bicycle and one amusing story concerned the mother of a young girl who was in service on the other side of Ipswich. As she had not seen her daughter for several weeks, she cycled from Framsden, near Debenham, on a Sunday afternoon against a strong head wind. When she arrived at her daughter's lodgings the girl could see that her mother was extremely tired. When asked what the trouble was her mother replied "Well dear, I've cycled all the way here against a strong wind so I hope it changes before I go back"

~ ~ ~ ~ ~ ~

During the last war a defused 1000 lb. bomb called "Hitler's Mistake" was paraded through Debenham and the sight of this prompted the elder Bloomfield sister, Marion, to set up a "Spitfire Fund" which eventually raised a very creditable sum of money.

~ ~ ~ ~ ~ ~

*Ernest W. List of Debenham. This bus was part of his Debenham & Ipswich Motor Service.*
Neil Bloomfield Collection.

A large bomb, or perhaps a land mine, fell and exploded very near to Church Farm, Ashfield, Malcolm Bloomfield's family home. His brother William, on hearing the explosion immediately took cover under the kitchen table and whilst lying there Willy heard his wife crawling about the kitchen floor. Being concerned for her safety he called and asked her what she was doing. "I'm trying to find my false teeth" she replied. Willy retorted "Don't be so silly, women, he's dropping bombs not fish and chips".

~ ~ ~ ~ ~ ~

W.G. Smoothy, a short time before his death on June 10th. 1908, ordered a steam motor tractor to be made embodying several alterations to the standard machine and these were incorporated. Unfortunately the machine was delivered on the day of his funeral. This tractor was exhibited at the 1908 Essex Agricultural Show and what happened to it next is not recorded.

~ ~ ~ ~ ~ ~

During the war a few farmers brought their threshing machines that required maintenance to Bloomfields using a pair of horses. On these occasions, having delivered the thresher, a motor-driven horsebox would be used to return the horses to the farm. On one occasion the horses were quite skittish and defied all the handlers attempts to get them up the ramp into the horse box. The church is very close to the Bloomfields Corner Works and perchance the vicar came by and asked the horseman if he could be of assistance. By now the man was becoming a bit waspy and he replied "Yes sir, you can tell me how Noah would have coaxed two cussed animals like these two into the Ark".

*The workers peace celebration, 1919. Harry Smoothy's wife, Ada, is fifth from the left in the back row. Oswold Smoothy's wife, Dorothy, is fourth from left in the back row* John Smoothy Collection.

~ ~ ~ ~ ~ ~

At Braiseworth Hall, Tannington, in the summer of 1946 there were a number of wagons, heavily loaded with hay, arriving at a large Dutch Barn, each to be backed-in for unloading by the Suffolk Punch that hauled it. One horse, handled by a young man, was obviously tired and just could not reverse the wagon into the barn. The late Reg Harvey, who was passing at the time, saw what was happening, took the reins and pressed his face close into the horses head. After a few moments the Suffolk Punch, without faltering, backed the wagon into the barn. Was Reg what is now known as a "Horse Whisperer"?.

*This pump carries the inscription "erected by voluntary subscription" and was in Rochford Market Square. When it was removed it laid in Smoothy's yard for many years and in the 1946 auction it was sold as scrap iron. To the right of the pump can be seen the shop of Goodman, Saddlers established in 1777.* John Smoothy Collection.

~ ~ ~ ~ ~ ~

Neil Bloomfield recounted a conversation between two women in the village as follows: One said "I'm sorry to hear about your husband passing away", to which the other replied "You know what a lazy devil he was, well I had him cremated and his ashes put into an egg timer. Now he has to work every time I boil an egg".

~ ~ ~ ~ ~ ~

John Smoothy believes that his grandfather owned the first pneumatic-tyred bicycle in Essex and his great-grandfather W.G. Smoothy had constructed, entirely of iron, one of the first bicycles seen in Chelmsford.

~ ~ ~ ~ ~ ~

Debenham Church was served to good effect by "Blind Harry" who pumped the organ and "Peg Leg" Hammond who maintained the large churchyard. "Blind Harry" was indeed blind and one of the legs of "Peg Leg" was a straight piece of wood, replacing the one he lost in World War I.

*Sir Robert Hitcham's School, Debenham, built c.1860. Keith Bloomfield has been a Trustee for about 34 years.*

Photo: Pat Freeman

~ ~ ~ ~ ~ ~

A collection, taken in Debenham in 1824, raised £110. 3s.0d. for the purpose of buying a Fire Engine for the village. It was supplied by Shand's of London for £97. It was a 16-man manual and was housed in a Station erected on Cross Green.

~ ~ ~ ~ ~ ~

Debenham Parish Council still owns a 4-wheeled hearse which was refurbished 30 years ago.

~ ~ ~ ~ ~ ~

A tradition of Bloomfields was an annual Christmas Eve get-together for all the staff which Mrs. Bloomfield masterminded. This took place by the forge, Mr. Muttick allowing his fire to be made up well after five o'clock. Alcoholic drink, mainly gin and beer, was served together with traditional mince pies and sausage rolls. With the affluence of the late 1950's, some of the staff acquired cars and motor cycles and being fearful that one of the youngsters might take a tumble from his motor cycle after drinking, Mrs.

Bloomfield decided that the tradition should cease. Since then a generous addition to the Christmas week pay packet has taken its place.

~ ~ ~ ~ ~ ~

A telephone call one Saturday in 1939 to Bloomfields (Tel. No. Debenham 53) from Mr. Harris, a Showman living near Clare, Suffolk, requested that someone be sent there to attend to leaking tubes in his Garrett Showmans engine. It had been sold to him a few years earlier by Malcolm for which he was paid with cash takings from the roundabout. Malcolm instructed Ernie Rozier to take his expanders and cycle over to Clare the following Monday morning to carry out the work. When out for a walk on the Sunday evening Malcolm met Ernie, also out for a walk with his wife, and stopped to offer him advice on how to get to Clare where the engine was situated. Ernie replied "That's alright Sir, I've been and had a look this afternoon and I know exactly where the engine stands". That was a round trip of about 75 miles by bicycle which he would repeat the next day! Talk about devotion to duty.

~ ~ ~ ~ ~ ~ ~ ~

One day Cecil Bloomfield came home with an amusing story. He had been called out to look at a binder by a small, not very efficient, farmer who declared that the binder and its tying mechanism was at fault because his machine did not deliver as many sheaves as his neighbours did. This neighbour was a good farmer who grew twice as much crop as did Cecil's customer. His sheaves were lying thick on the ground, whereas this poor man's were several yards apart. There was nothing wrong with his binder but he would not accept Cecil's explanation that it was the crop that was at fault not the binder.

*Debenham High Street looking north. The building on the right is Dove Cottage, owned by the Church, in which the Parish Council hold their meetings.*

Photo: Pat Freeman